Springer Series on
Atoms+Plasmas

9

Editor: G. Ecker

Springer Series on

Atoms+Plasmas

Editors: G. Ecker   P. Lambropoulos   I. I. Sobel'man   H. Walther
Managing Editor: H. K. V. Lotsch

E. Oks

Plasma Spectroscopy

The Influence of Microwave and Laser Fields

With 74 Figures and 3 Tables

Springer-Verlag

Berlin Heidelberg New York London Paris
Tokyo Hong Kong Barcelona Budapest

Sep/ae
Phys

Prof. Dr. Eugene Oks
Auburn University
Dept. of Physics
206 Allison Laboratory
Auburn, AL 36849-5311
USA

Series Editors:

Professor Dr. Günter Ecker
Ruhr-Universität Bochum, Lehrstuhl Theoretische Physik I, Universitätsstrasse 150,
D-44801 Bochum, Germany

Professor Peter Lambropoulos, Ph.D.
Max-Planck-Institut für Quantenoptik
D-85748 Garching, Germany, and
Foundation of Research and Technology – Hellas (FO.R.T.H.)
Institute of Electronic Structure and Laser (IESL) and
University of Crete, PO Box 1527, Heraklion, Crete 71110, Greece

Professor Igor I. Sobel'man
Lebedev Physical Institute, Russian Academy of Sciences,
Leninsky Prospekt 53, 117924 Moscow, Russia

Professor Dr. Herbert Walther
Universität München, Sektion Physik, Am Coulombwall 1,
D-85748 Garching/München, Germany

Managing Editor: Dr. Helmut K.V. Lotsch
Springer-Verlag, Tiergartenstrasse 17, D-69121 Heidelberg, Germany

ISBN 3-540-54100-4 Springer-Verlag Berlin Heidelberg New York
ISBN 0-387-54100-4 Springer-Verlag New York Berlin Heidelberg

Library of Congress Cataloging-in-Publication Data. Oks, E.A. (Evgeniĭ Aleksandrovich) Plasma spectroscopy: the influence of microwave and laser fields/Eugene Oks. p. cm. – (Springer series on atoms and plasmas; 9) Includes bibliographical references (p.). ISBN 0-387-54100-4 (alk. paper). 1. Plasma spectroscopy. 2. Plasma diagnostics. 3. Microwave plasmas. 4. Laser beams. I. Title. II. Series. QC718.5.S6037 1995 530.4\46 – dc20 94-35448

Typesetting: Macmillan India Ltd., Bangalore-25

SPIN: 10017902   31/3145/SPS–5 4 3 2 1 0–Printed on acid-free paper

Dedicated to My Parents

Preface

Questions raised in various areas of applied plasma research motivated the development of spectroscopic diagnostics for systems in which strong monochromatic electric fields or quasimonochromatic electric fields (QEFs) are important.

In the course of time, the focus in studying plasma instabilities has shifted from turbulent, broadband electric fields to QEFs. The reason is that longitudinal QEFs may be excited, for example in pulsed discharges, which are employed as advanced sources of powerful neutron and X-ray radiation, or during interaction of a plasma with high-current beams of charged particles. Even more interesting are transverse QEFs, for example, laser or maser radiation, which can penetrate a plasma from the exterior or can be generated in a plasma. The transverse QEFs determine to a significant extent physical processes in microwave heating of plasmas in tokamaks, in laser fusion, and in technological microwave discharges (plasma processing) and are of major importance for investigations on plasma lasers and radiowave propagation through the ionosphere.

For these reasons the development of non-perturbing spectroscopic methods for the diagnostics of plasma media containing QEFs has become an urgent task. For theoreticians this represents a new class of plasma-spectrosopic problems – the radiation of a quantum system interacting simultaneously with an intense QEF and a plasma medium. This monograph is devoted to this theoretical investigation, in particular the elaboration of new methods for plasma diagnostics and their practical utilization.

In the Introduction (Chap. 1) the problems are set up and the boundaries of the subject and its interconnection with other research areas are specified. Chapter 2 reviews known relevant theoretical-analytical methods as well as the latest results not previously published. The content of this chapter has a more general quantum-mechanical character and is of importance not only for the spectroscopy of plasmas with QEFs but also for nonlinear optics. (Chapter 2 may be skipped without affecting comprehension of the rest of the book.) Chapters 3–6 discuss the main spectroscopic effects of the interaction of radiating atoms and ions with QEFs and other electric fields caused by plasma ions and electrons. In Chap. 7 I present practical applications of these results to experiments in various areas of applied plasma research.

I am very grateful to Prof. Dr. V.S. Lisitsa, with whom I discussed some of the problems. In writing the theoretical sections I have used some results of investigations carried out jointly with my former and current graduate students Dr. V.P. Gavrilenko, Dr. B.B. Nadezhdin, D.A. Volod'ko, Ya.O. Ispolatov, and A.P. Derevianko as well as with my former colleague Dr. I.M. Gaisinsky. I am thankful to all of them.

This monograph was written mostly during my stay at the Institute of Experimental Physics V of the Ruhr University at Bochum, Germany. I owe much to the Alexander von Humboldt Foundation, which sponsored my stay. I also wish to express my gratitude to Prof. Dr. H.J. Kunze, Director of the Institute, for his wonderful hospitality and for giving me an opportunity to write this book. I am especially grateful to Prof. Dr. G. Ecker, the responsible series editor, for his valuable comments. I thank Dr. S. Maurmann for his help and express my gratitude to Mrs. K. Waldenburg, who has patiently typed this manuscript.

Auburn, USA   E. Oks
November 1994

Contents

List of Abbreviations

BEF   broadband electric field
BLR   broadband laser radiation
DPM   diatomic polar molecule
ED   ellipticity degree
EF   electric field
EOS   electron oscillatory shift
ICL   intracativity laser
MES   molecular emission spectra
PT   perturbation theory
QEF   quasimonochromatic electric field
QS   quasienergy state
CSZO   correct states of zeroth order
RTA   rectilinear trajectories approximation
SL   spectral line
TS   thomson scattering
WF   wave function

1 Introduction

In both laboratory and natural plasmas there are two classes of Electric Fields
(EFs), distinguished by the relative width of the frequency band $\delta\omega/\omega$. These are
the Quasimonochromatic Electric Fields (QEFs, $\delta\omega/\omega \ll 1$) and the Broadband
Electric Fields (BEFs, $\delta\omega/\omega \gtrsim 1$), which act quite differently on radiating
quantum systems (radiators or emitters). A plasma containing no QEFs can be
seen, from the spectroscopic point of view, as BEFs of several frequency ranges
acting on a radiator. That is because the chaotic thermal motion of electrons
and ions corresponds to individual BEFs, whereas collective BEFs can arise as
a result of plasma turbulence. To this collective BEF belong, for example, the
oscillations which may be responsible for the anomalous resistivity of a plasma:
ionic sound waves, Bernstein modes, etc. [1.1]. In the reference frame of ions
these oscillations are of low frequency, concentrated in the band $(0, \omega_{pi})$, where
$\omega_{pi} = (4\pi e N_e/M)^{1/2}$ is the ionic plasma frequency (N_e is the electron density).

The theoretical explanation of spectral line broadening by individual BEFs
has been developed in detail. It forms the basis of various spectroscopic diagnostic methods for the determination of N_e as well as the electron T_e and ion
T_i temperatures [1.2]. The development of analogous methods for the measurements of parameters of collective BEFs started over 20 years ago [1.3]. In
principle, the influence of individual and collective BEFs on the radiator do
not differ. Moreover, because of their low frequency (electron individual BEFs
excepted) the action of BEFs may usually be treated quasistatically.

It should be noted that other static or quasistatic EFs may also be present
in a plasma, especially in tokamak plasmas. For example, electrically biased
structures in a plasma affect the potential profile and the resulting EF strongly
influences transport phenomena. Or, when neutral beams are injected into
a plasma across a magnetic field B they "feel" the effective Lorentz field
$F = v \times B/c$ (in present-day experiments – of order 10–100 kV/cm) [1.4].

It is important that both quasistatic BEFs and static EFs may be described
in terms of a static vector field F characterized by an ensemble distribution
$W(F)$. All these fields produce no dynamic spectroscopic phenomena.

However, the presence of QEFs in a plasma may change this situation fundamentally: the time evolution of a radiator may acquire a dynamic character in spite of relaxation processes so that the averaged motion of an optical electron may be described in terms of precession, nutation, etc. This leads to the appearance of new components in the radiation spectra: in the simplest cases the satellites occur at distances which are a multiple of the QEF frequency ω.

The further details of the physical situation may be outlined as follows. The energy spectrum of a radiator (atom, ion) in a plasma consists of levels (multiplets) comparatively distant from each other, each possessing a microstructure with a characteristic scale $\Delta \ll \omega_0$ where ω_0 is the separation between the multiplets. For the QEF frequency ω the inequality $\omega \ll \omega_0$ holds, but generally no restrictions are imposed on the ratio ω/Δ. As a result of the interaction of three subsystems – radiator (R), QEF (F) and plasma medium (P) represented, e.g. by a quasistatic EF – the microstructure of levels of an emitter may be substantially modified. The observed spectrum corresponds to radiative transitions (spontaneous, usually) at a frequency of order ω_0. The radiation is a weak probe signal that reflects the perturbed level microstructure and carries out from the plasma this information about the parameters of the QEF and plasma medium.

It should be emphasized that for the overwhelming majority of practical problems of QEF diagnostics in plasmas the QEF frequency is really much smaller than the frequency of the observed Spectral Line (SL): $\omega \ll \omega_0$. This is valid not only for low-temperature plasmas, in which neutral atoms emit in the visible range, but also for high-temperature plasmas, where multicharged ions emit in the X-ray range. When the QEF is a laser field whose frequency is close to an atomic transition frequency $|\omega_0 - \omega|/\omega_0 \ll 1$ then nonlinear effects can occur. Their relation to SL broadening theory has been investigated [1.5].

Chapter 2 is devoted to further development of the analytical methods for describing the interaction of quantum systems with a nonstationary field. For a very wide class of quantum-mechanical problems in which a quantum system interacts with an external field periodic in time, the formalism of quasienergy states (QSs) seems to be most suitable. The terminology "quasienergy state" and "quasienergy" was introduced in [1.6, 7].

The QSs are defined as follows. Let the Hamiltonian of a quantum system be time-periodic: $H(x, t + T) = H(x, t)$. Then from the solutions $\psi(x, t)$ of the Schrödinger equation (in the system of units with $\hslash = 1$)

$$i\, \partial\psi/\partial t = H(t)\psi, \tag{1.1}$$

the following particular solutions of QS type may be singled out:

$$\psi_n(x, t) = \exp(-i\varepsilon_n t)\varphi_n(x, t), \tag{1.2}$$

where $\varphi_n(x, t)$ is a time-periodic function: $\varphi_n(x, t+T) = \varphi_n(x, t)$. A quantum state of type $\psi_n(x, t)$ is called a QS and the quantity ε_n is called the quasienergy of this state.

We now determine the formula for the radiation spectrum corresponding to the transition of a quantum system between the QSs $\psi_n(x, t)$ and $\psi_m(x, t)$.

The starting expression for the SL profile $I_{nm}(\Delta\omega)$ is [1.8]

$$I_{nm}(\Delta\omega) \propto \frac{1}{2\pi\tau} \left| \int_{-\tau/2}^{\tau/2} \exp(-it\,\Delta\omega)\langle\psi_n(x,t)|d(x)|\psi_m(x,t)\rangle\,dt \right|^2, \qquad (1.3)$$

where $d(x)$ is the operator connecting the QS; for example, $d = er$ in the dipole approximation for an atomic electron. Let us expand the periodic parts of the $\varphi_{n,m}(x,t)$ QS in a Fourier series:

$$\varphi_{n,m}(x,t) = \sum_{j=-\infty}^{+\infty} C_j^{n,m}(x)\,\exp(-ij\omega t), \qquad \omega \equiv 2\pi/T. \qquad (1.4)$$

Substituting (1.4) into (1.3) and singling out the δ-function according to the formula

$$\lim_{\tau\to\infty} \frac{1}{\tau} \int_{-\tau/2}^{\tau/2} dt\,\exp[-i(\Delta\omega-\varepsilon)t] \propto \delta(\Delta\omega-\varepsilon),$$

we obtain the final expression

$$I_{nm}(\Delta\omega) \propto \sum_{j=-\infty}^{+\infty} \left| \sum_{k=-\infty}^{+\infty} \langle C_{j+k}^n(x)|d(x)|C_k^m(x)\rangle \right|^2 \delta(\Delta\omega - (\varepsilon_n - \varepsilon_m + j\omega)).$$

$$(1.5)$$

Thus the spectrum of spontaneous emission for a quantum system in a time-periodic field consists of the set of satellites at frequencies $\Delta\omega = \varepsilon_n - \varepsilon_m + j\omega\,(j = 0,\pm1,\pm2,\dots)$, where ε_n and ε_m are the quasienergies of the various QSs of the system in this field. In Chap. 2 special attention will be paid to quite new general theoretical results for QSs which have not been published yet.

In Chaps. 3–6 we present radiation spectra of various quantum systems under the action of QEF or under the joint action of QEF and quasistatic EF calculated analytically. In Chaps. 3 and 4 the "Coulomb radiator" is considered. This is a hydrogen-like emitter in the nonrelativistic approximation. All other quantum systems are labeled as "non-Coulomb radiators". Their radiation spectra, calculated for analogous conditions, are presented in Chaps. 5 and 6.

Chapters 3–6 discuss a series of new phenomena in plasma spectroscopy (some of them were briefly described in our short reviews [1.9, 10]. If the BEF influence on a radiator has predominantly quasistatic character then it can tune the microstructure of radiator energy levels to a resonance with QEF. Individual radiators feel, generally speaking, different strengths of the quasistatic EF F that is described by some distribution function $W(F)$. That is why for a given QEF frequency ω one can single out from the ensemble of radiators one or more groups of atoms (or ions) which experience the action of the field F just corresponding to the conditions of the resonance with the QEF. The detailed investigations of this effect are now grouped together as intra-Stark spectroscopy, that is, spectroscopy inside the static Stark profile. For certain nonresonant cases even relatively weak quasistatic EFs may drastically change the result of the QEF action on a radiator. Interesting effects also occur under the joint action of a quasistatic EF F and high-frequency nonresonant QEF. This dynamic problem may be reduced to a static one with a redefined static EF F_{eff} depending on the parameters of the QEF. Since the inequality $F_{\text{eff}} < F$ holds, the effect might be interpreted as a partial suppression of the quasistatic EF action on a radiator by the high-frequency QEF.

If the BEF influence on a radiator has predominantly impact character, the situation further depends on whether the QEF or BEF has the more rapid influence. In the first case the impact action of the BEF is primarily felt by the QS of the radiator, their impact width becoming a complicated oscillating function of the amplitude and the frequency of QEF. In the second case during the time of formation of impact broadening the QEF practically does not change. Nevertheless, the QEF leads to an anisotropy in the velocity distribution of the perturbing particles and also to a change in the microstructure of radiator energy levels. As a result, a significant impact shift arises and also the impact width decreases, both effects being nonlinearly connected with the QEF energy density.

Spectra of laser absorption or fluorescence of atoms and molecules in plasmas containing QEFs also show innovative features. The most intriguing effects arise under the joint action of a strong QEF and a laser field which is resonant to some transition between atomic multiplets. Not only the enumerated physical effects but also the ensueing methods proposed for QEF diagnostics in low- and high-temperature plasmas are quite new. The results of their experimental application are described in Chap. 7.

We emphasize that it was the aim of the theoretical calculations to find the dependence of the emitted SL profiles $I(\Delta\omega)$ on the QEF parameters such as the amplitude E_0, the peak position ω, and the width $\delta\omega$ of the frequency spectrum, the ellipticity degree ξ, and the spatial orientation, as well as on the plasma medium parameters like the density N_e and the temperature T_e of electrons, the temperatures of ions T_i and atoms T_a. However, in practice, application of the theory of Chaps. 3–6 requires methods of solving the "inverse" problem: the determination of the parameters of the QEF and the plasma from observed SL profiles. Since this inverse problem is much more complicated than the "direct" one, unique approaches have to be found for each experiment. Chapter 7 demonstrates the hidden problems which accompany this application of the theoretical results.

This book is written as an advanced-level monograph on shapes and shifts of spectral lines in plasmas. Therefore, the author refrained from reproducing the fundamentals of quantum mechanics or traditional theory of spectral line broadening in plasmas which are necessary for understanding the material. This seems permissible in light of the availability of well-known textbooks and monographs on these topics, e.g. [1.2, 8, 11].

The following notations accepted in plasma spectroscopy are used: $\Delta\lambda$, the detuning relative to the unperturbed wavelength λ_0 of the SL; $\Delta\omega$, the detuning in the frequency scale relative to the unperturbed frequency ω_0 of SL; indices α, α', α'', ... correspond to states of the upper multiplet, indices β, β', β'', ... correspond to states of the lower multiplet, the SL arises as a result of radiative transitions between these states; (n, n_1, n_2, m) parabolic quantum numbers, (n, l, m) spherical quantum numbers, (n, l, j, m_j) quantum numbers of states described by Pauli wave functions (WFs). In order to distinguish quantum numbers of upper states from quantum numbers of lower states the latter are labeled by a prime. The following abbreviations are used: EF electric field, QEF quasimonochromatic electric field, BEF broadband electric field, SL spectral line, QS quasienergy state, WF wave function.

2 Analytical Methods for the Calculation of Quasienergy States of Quantum Systems

This chapter develops analytical methods for describing an interaction of quantum systems with a nonstationary field. First, a general review of relevant methods is given. Then new (previously unpublished) analytical methods (perturbational and nonperturbational) for the calculations of QS are presented.[1]

2.1 Interaction of Quantum Systems with a Nonstationary Field

We shall single out from the stationary energy levels of a quantum system a group of close-lying levels with energies ε_k separated by the value $|\varepsilon_k - \varepsilon_{k'}| \sim \Delta$ (here and below $\hbar = 1$). The other levels of the quantum system are separated from this group by a value of order $D \gg \Delta$. (This is the case, e.g. for atomic multiplets.) The time-periodic interaction $V(x, t + 2\pi/\omega) = V(x, t)$ with an external field mixes (generally strongly) the states of the considered group. No initial restrictions are imposed on the ratios V/Δ and ω/Δ. It is only assumed that $D \gg \max(V, \omega)$ so that the mixing of the states from the considered group with distant states may be allowed for, if necessary, by the perturbation theory (PT). The observed SL with frequency ω_0 corresponds to radiative transitions between the considered group of levels and some of the distant levels: $\omega_0 \gtrsim D$.

The simplest analytical method for the treatment of such nonstationary problems is the well-known *Dirac' PT* merely called "nonstationary PT" [2.1–3]. For the interaction of type $V(x) \cos \omega t$, the quasienergy $\tilde{\varepsilon}_k$ of an isolated state with energy ε_k is given (in the second order of perturbation theory) by [2.4]

$$\tilde{\varepsilon}_k = \varepsilon_k + 2^{-1} \sum_i \omega_{ki} |\langle i|V(x)|k\rangle|^2 / (\omega_{ki}^2 - \omega^2), \quad \omega_{ki} \equiv \varepsilon_k - \varepsilon_i, \quad (2.1.1)$$

where the summation extends over all other unperturbed states $|i\rangle$ of the quantum system. The condition of validity generally may be expressed as $V \ll |\Delta^2 - \omega^2|^{1/2}$. It should be mentioned, however, that for nonisolated, in particular, degenerate states interacting with a time-periodic field the PT of QS was developed only recently, the results are given in Sect. 2.2.

---

[1] These results were obtained by *B.B. Nadezhdin* in the course of his doctoral studies under the author's guidance.

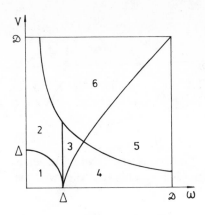

Fig. 2.1. The regions of applicability of analytical methods describing the interaction of quantum systems with a nonstationary field. The potential V in the Hamiltonian is chosen as the ordinate, the frequency of the field ω-as the abscissa. The characteristic separation between the levels is $\Delta \ll D$, where D is the minimum distance to other levels of the quantum system. Dirac perturbation theory – regions 1, 4, 5; the adiabatic approximation – regions 1, 2, 3, 4; high-frequency approximation – regions 3, 4, 5, 6. In narrow bands (not shown in the figure) near the lines $q\omega = \omega_{ij}^{(0)}$ ($\omega_{ij}^{(0)}$ is the separation between the levels i and j) the resonance approximation may also be applied

In the case in which the distance ω_{ki} between two levels is close to a multiple of the external field frequency ($\omega_{ik} \approx qw$, $q = 1, 2, 3, \ldots$) the familiar *resonance approximation* may be applied [2.1, 3]. The validity condition is $|\omega_{ik} - qw| << \omega_{ik}$. The restriction on V depends upon q; e.g. for $q = 1$ it must be $V \ll \omega_{ik}$.

In the case of a low frequency external field the well-known *adiabatic approximation* may be used [2.2, 3]. The validity condition may be expressed as $\omega V \ll \Delta^2$. Thence it is clear that for $\omega \ll \Delta$ the adiabatic approximation allows one to advance to the strong field region $V \gtrsim \Delta$ for which PT is invalid.

The opposite case of a high frequency external field was until recently not as well understood as the previous cases. The development of a *high-frequency approximation* in a general (but explicit) form is presented in Sect. 2.3. The condition of validity is $\max(\omega V/\Delta^2, \omega^2/\Delta^2) \gg 1$. This means that the field must be either of high frequency or strong.

The regions of applicability of the above approximations have some overlap. This can be easily seen from Fig. 2.1, in which these regions are shown.

2.2 Perturbation Theory for QSs of Degenerate Quantum Systems

2.2.1 Calculation of the QSs as a Stationary Problem

It is known that the calculation of QSs for a quantum system with a *stationary* unperturbed Hamiltonian $H(x)$ and time-periodic perturbation $V(x, t)$ may be reduced to a stationary problem [2.5, 6]. We shall develop an algorithm for a more general problem with a *nonstationary* unperturbed Hamiltonian $H(x, t)$.

Consider a quantum system with a time-periodic unperturbed Hamiltonian $H(x, t + T) = H(x, T)$. *Zel'dovich* [2.7] has shown that for such $H(x, t)$ the set of QSs

$$\Psi_{n\alpha}(x, t) = \exp(-i\varepsilon_n t)\varphi_{n\alpha};$$
$$\varphi_{n\alpha}(x, t) = \varphi_{n\alpha}(x, t + T), \qquad \langle\varphi_{n\alpha}|\varphi_{m\beta}\rangle = \delta_{nm}\delta_{\alpha\beta} \qquad (2.2.1)$$

exists. Here subscripts α, β are used to allow for a possible degeneracy of the level ε_n; δ_{nm} is the Kronecker symbol. The functions $\varphi_{n\alpha}(x, t)$ satisfy the equation

$$(H - i\,\partial/\partial t)\varphi_{n\alpha} = \varepsilon_n\varphi_{n\alpha}. \qquad (2.2.2)$$

We introduce the Hilbert space $L_2(x \otimes t)$ of functions $\varphi(x, t)$ which depends on the coordinates x of the configuration space (and are time-periodic) with the scalar product

$$\langle\langle\varphi_1|\varphi_2\rangle\rangle \equiv \frac{1}{T} \int_0^T dt\,\langle\varphi_1(t)|\varphi_2(t)\rangle. \qquad (2.2.3)$$

From (2.2.2) it is seen that $\varphi_{n\alpha}(x, t)$ and ε_n will be, respectively the eigenfunctions and the eigenvalues of $H - i\,\partial/\partial t$, which is a Hermitian operator in $L_2(x \otimes t)$ [2.6]. It is obvious that any function $\varphi_{n\alpha}(x, t)\exp(iN\omega t)$ ($\omega = 2\pi/T$, $N = 0, \pm 1, \ldots$) will also be an eigenfunction of $H - i\,\partial/\partial t$ with the eigenvalue $\varepsilon_n + N\omega$. As all such functions are time-periodic with the period $T = 2\pi/\omega$, the set of eigenfunctions of $H - i\,\partial/\partial t$ in $L_2(x \otimes t)$ has the form

$$|Nn\alpha\rangle\rangle = \varphi_{n\alpha}(x, t)\exp(iN\omega t), \qquad N = 0, \pm 1, \pm 2, \ldots;$$
$$(H - i\,\partial/\partial t)|Nn\alpha\rangle\rangle = (\varepsilon_n + N\omega)|Nn\alpha\rangle\rangle,$$
$$\langle\langle Nn\alpha|N'n'\alpha'\rangle\rangle = \delta_{NN'}\delta_{nn'}\delta_{\alpha\alpha'} \qquad (2.2.4)^2$$

Now let the quantum system experience a time-periodic perturbation $\lambda V(x, t) = \lambda V(x, t + T)$. We try to find solutions $\tilde{\Psi}(x, t)$ of the Schrödinger equation $i\,\partial\tilde{\Psi}/\partial t = \tilde{H}\tilde{\Psi}$, $\tilde{H} = H + \lambda V$ in the form of QSs

$$\tilde{\Psi}_{n\alpha}(x, t) = \exp(-i\tilde{\varepsilon}_{n\alpha}t)\tilde{\varphi}_{n\alpha}(x, t). \qquad (2.2.5)$$

[The quasienergy $\tilde{\varepsilon}_{n\alpha}$ in (2.2.5) has the index α, since under the action of the perturbation $\lambda V(x, t)$ the splitting of degenerate quasienergy levels ε_n may occur.]

By analogy with (2.2.2) the periodic functions $\tilde{\varphi}_{n\alpha}(x, t)$ in (2.2.5) are the eigenfunctions of $\tilde{H} - i\,\partial/\partial t = H + \lambda V - i\,\partial/\partial t$ in $L_2(x \otimes t)$ with the eigenvalues $\tilde{\varepsilon}_{n\alpha}$:

$$(H - \lambda V - i\partial/\partial t)\tilde{\varphi}_{n\alpha} = \tilde{\varepsilon}_{n\alpha}\tilde{\varphi}_{n\alpha}. \qquad (2.2.6)$$

We expand the states $\tilde{\varphi}_{n\alpha}$ in the known set of QSs (2.2.4):

$$\tilde{\varphi}_{n\alpha}(x, t) = \sum_{M,m,\beta} C_{n\alpha}^{Mm\beta}|Mm\beta\rangle\rangle. \qquad (2.2.7)$$

---

^{2} In this section, capital Latin letters denote QS harmonics ($N\omega$), lower case Latin letters quasienergies (ε_n), and Greek letters degenerate states ($\varphi_{n\alpha}$).

Substituting (2.2.7) into (2.2.6), taking into account the action of $H - i\,\partial/\partial t$ on the states $|Mm\beta\rangle\rangle$ of the type (2.2.4) and calculating the scalar product of the obtained equation with an arbitrary vector $\langle\langle Ll\gamma|$, we get the usual system of linear equations for determining $C_{n\alpha}^{Mm\beta}$ and $\tilde{\varepsilon}_{n\alpha}$

$$\sum_{M,m,\beta} \left\{ C_{n\alpha}^{Mm\beta} [\lambda\langle\langle Ll\gamma|V|Mm\beta\rangle\rangle + \delta_{ML}\delta_{ml}\delta_{\beta\gamma}(\varepsilon_l + L\omega)] \right\} = \tilde{\varepsilon}_{n\alpha} C_{n\alpha}^{Ll\gamma},$$

$$(2.2.8)$$

where

$$\langle\langle Ll\gamma|V|Mm\beta\rangle\rangle \equiv \frac{1}{T} \int\limits_{0}^{T} \exp[i(M-L)\omega t] \int dx\, \varphi_{l\gamma}^*(x,t) V(x,t) \varphi_{m\beta}(x,t).$$

$$(2.2.9)$$

The matrix elements (2.2.9) are Fourier components $V^{(M-L)}$ of time-dependent matrix elements of $V(x,t)$ calculated on QSs $\varphi_{n\alpha}(x,t)$.

From the system (2.2.8) it is seen that the *sought quasienergies* $\tilde{\varepsilon}_{n\alpha}$ of the perturbed Hamiltonian $\tilde{H} = H + \lambda V$ and the *desired expansion coefficients* $C_{n\alpha}^{Mm\beta}$ represent respectively, *the eigenvalues and the eigenvectors of the matrix* $\tilde{H}_{Ll\gamma}^{Mm\beta}$, which is constructed from the set of QSs (2.2.4):

$$\tilde{H}_{Ll\gamma}^{Mm\beta} \equiv \lambda\langle\langle Ll\gamma|V|Mm\beta\rangle\rangle + \delta_{ML}\delta_{ml}\delta_{\beta\gamma}(\varepsilon_l + L\omega).$$

$$(2.2.10)$$

Thus the nonstationary problem of the QS calculation is reduced to a stationary one, namely, to a diagonalization of the matrix $\tilde{H}_{Ll\gamma}^{Mm\beta}$.

2.2.2 Perturbation Theory

The PT for the calculations of quasienergies $\tilde{\varepsilon}_{n\alpha}$ and WFs $\tilde{\varphi}_{n\alpha}$ (2.2.6) may be constructed by analogy to the stationary PT. It is necessary to solve the following two problems.

1. If the unperturbed levels $\varepsilon_n + N\omega$ are degenerate, the so-called "correct states in zeroth order" (CSZO) $\varphi_{n\alpha}$ must be found [2.1]. Note that the case of *resonant* perturbation ($\varepsilon_n + N\omega = \varepsilon_{n'} + N'\omega$ for some levels) is described in the language of initial states $|Nn\alpha\rangle\rangle$ just as a *QS degeneracy* [2.6].

2. The corrections to ε_n and $\varphi_{n\alpha}$ must be calculated. We try to find $\tilde{\varepsilon}_{n\alpha}$ and $\tilde{\varphi}_{n\alpha}$ in the form of a series in powers of λ:

$$\tilde{\varepsilon}_{n\alpha} = \varepsilon_n + \lambda\varepsilon_{n\alpha}^{(1)} + \lambda^2\varepsilon_{n\alpha}^{(2)} + \cdots; \qquad \tilde{\varphi}_{n\alpha} = \sum_{M,m,\beta} C_{n\alpha}^{Mm\beta}|Mm\beta\rangle\rangle,$$

$$C_{n\alpha}^{Mm\beta} = \delta_{M0}\delta_{mn}\delta_{\alpha\beta} + \lambda A_{n\alpha}^{Mm\beta} + \lambda^2 B_{n\alpha}^{Mm\beta} + \cdots.$$

$$(2.2.11)$$

In the zeroth order of the $C_{n\alpha}^{Mm\beta}$ expansion, $C_{n\alpha}^{Mm\beta} = \delta_{M0}\delta_{mn}\delta_{\alpha\beta}$ is set up, i.e. $\tilde{\varphi}_{n\alpha}^{(0)} = |0n\alpha\rangle\rangle = \varphi_{n\alpha}$. In other words we have demanded $\varphi_{n\alpha}$ to be CSZO. Substituting (2.2.11) into (2.2.8) and writing out the terms of the same order of λ, we obtain the equations of PT.

The zeroth order: for all L, l, γ we have

$$\delta_{L0}\delta_{ln}\delta_{\alpha\gamma}(\varepsilon_l + L\omega) = \varepsilon_n\delta_{L0}\delta_{ln}\delta_{\alpha\gamma}.$$

The first order:

$$\langle\langle Ll\gamma|V|0n\alpha\rangle\rangle + A_{n\alpha}^{Ll\gamma}(\varepsilon_l + L\omega) = \varepsilon_n A_{n\alpha}^{Ll\gamma} + \varepsilon_{n\alpha}^{(1)}\delta_{L0}\delta_{ln}\delta_{\alpha\gamma} \qquad (2.2.12)$$

for all L, l, γ. Substituting $L = 0$, $l = n$ in (2.2.12) we obtain the *secular equation* for determining the CSZO

$$\varepsilon_{n\alpha}^{(1)}\delta_{\alpha\gamma} = \langle\langle 0n\gamma|V|0n\alpha\rangle\rangle \equiv \frac{1}{T}\int\limits_0^T dt\int dx\,\varphi_{n\gamma}^*(x,t)\varphi_{n\alpha}(x,t). \qquad (2.2.13)$$

Thus the CSZO are chosen so as to diagonalize the perturbation matrix $V(x,t)$ in (2.2.13) (averaged over the period T) constructed on all QSs $\varphi_{n\alpha}$, $\varphi_{n\gamma}$ of the degenerate level ε_n. The eigenvalues of this matrix are the first-order corrections $\varepsilon_{n\alpha}^{(1)}$ to the quasienergies.

Substituting $l \neq n$ or $L \neq 0$ in (2.2.12) we find the formulas for the first order of expansion coefficients:

$$A_{n\alpha}^{Ll\gamma} = \langle\langle Ll\gamma|V|0n\alpha\rangle\rangle(\varepsilon_n - \varepsilon_l - L\omega)^{-1} \quad (1 \neq n \text{ or } L \neq 0). \qquad (2.2.14)$$

As the resonant states are considered to be degenerate (see above) the denominators in (2.2.14) cannot be equal to zero for $l \neq n$.

The second order: If all values $\langle\langle 0n\gamma|V|0n\alpha\rangle\rangle = 0$ for QSs belonging to the same degenerate level ε_n, then we do not obtain the secular equation (2.2.13) for determining the CSZO. In this case $\varepsilon_{n\alpha}^{(1)} = 0$ and for the second order terms from (2.2.11) and (2.2.8) we get (for all L, l, γ)

$$\sum_{M,m,\beta} A_{n\alpha}^{Mm\beta}\langle\langle Ll\gamma|V|Mm\beta\rangle\rangle + B_{n\alpha}^{Ll\gamma}\cdot(\varepsilon_l + L\omega) = \delta_{L0}\delta_{ln}\delta_{\alpha\gamma}\varepsilon_{n\alpha}^{(2)} + \varepsilon_n B_{n\alpha}^{Ll\gamma}.$$

$$(2.2.15)$$

Replacing $L = 0, l = n$ in (2.2.15) and substituting $A_{n\alpha}^{Mm\beta}$ from (2.2.14) we obtain the *secular equation* for determining the CSZO (for the case $\varepsilon_{n\alpha}^{(1)} = 0$):

$$\varepsilon_{n\alpha}^{(2)}\delta_{\alpha\gamma} = \sum_{M,m,\beta}\langle\langle 0n\gamma|V|Mm\beta\rangle\rangle\langle\langle Mm\beta|V|0n\alpha\rangle\rangle(\varepsilon_n - \varepsilon_m - M\omega)^{-1} \equiv V_{\alpha,\gamma}^n.$$

$$(2.2.16)$$

Thus in the case $\varepsilon_{n\alpha}^{(1)} = 0$ the CSZO are chosen so as to diagonalize the matrix $V_{\alpha,\gamma}^n$, where (2.2.16) is constructed on all QSs of the degenerate level ε_n. The eigenvalues of this matrix represent the second-order corrections $\varepsilon_{n\alpha}^{(2)}$ to the quasienergies.

The criterion of applicability of PT is reduced (as in the stationary case) to the requirement that the corrections to quasienergies are small compared with the separation between the unperturbed quasienergies $|(\varepsilon_n + N\omega) - (\varepsilon_m + M\omega)|$.

2.3 High-Frequency or Very Intense Nonstationary Fields

2.3.1 Calculation of the QSs as a Stationary Problem

Let an unperturbed *stationary* Hamiltonian $H(x)$ have a set of discrete eigenstates $\varphi_n(x) \equiv |n\rangle$: $H|n\rangle = \varepsilon_n|n\rangle$, ε_n is the unperturbed energy of the state $|n\rangle$, $\langle m|n\rangle = \delta_{mn}$ (index α, which designated a degeneracy in (2.2.1), does not appear here since for this formalism the degeneracy of the levels ε_n is not important). Let the states $|n\rangle$ experience the action of a strong time-periodic field which corresponds in the Hamiltonian to the following term:

$$\lambda V(x, t) = \lambda U(x) f(t), \qquad f(t + T) = f(t) \tag{2.3.1}$$

[$V(x)$ is a Hermitian operator in the space of functions $|n\rangle$, $f(t)$ is a real function].

If we calculate the stationary matrix $\langle n|U(x)m\rangle$, it will be, generally speaking, nondiagonal. We designate by Greek letters $|\alpha\rangle$, $|\beta\rangle$ such linear combinations of eigenstates of the Hamiltonian H which diagonalize the U-matrix

$$\langle \alpha|U(x)|\beta\rangle = U_\alpha \delta_{\alpha\beta}, \qquad \langle \alpha|\beta\rangle = \delta_{\alpha\beta}. \tag{2.3.2}$$

In practice, to find the states $|\alpha\rangle$ it is necessary to find the eigenvalues U_α of the matrix $\langle n|U(x)|m\rangle$ and the corresponding eigenvectors C_n^α, so that

$$|\alpha\rangle = \sum_n C_n^\alpha |n\rangle. \tag{2.3.3}$$

First we shall construct a set of QSs of the type (2.2.4) for the operator $\lambda V - i\,\partial/\partial t$ ($H = 0$) and then we shall seek in the basis of these QSs the eigenstates for the operator $H + \lambda V - i\,\partial/\partial t$. Physically this means that for very strong fields ($\lambda \to \infty$) the initial Hamiltonian H should be treated as a perturbation.

From the action of $U(x)$ on the states $|\alpha\rangle$ given by (2.3.3) it is easy to see that the WFs

$$\Psi_\alpha(x, t) = |\alpha\rangle \exp\left(-i\lambda U_\alpha \int_{t_0}^t f(\tau)\,d\tau\right) \tag{2.3.4}$$

are the solutions of the Schrödinger equation

$$i\,\partial\Psi_\alpha/\partial t = \lambda V(x, t)\Psi_\alpha, \qquad \lambda V(x, t) = \lambda U(x) f(t). \tag{2.3.5}$$

The WFs (2.3.4) are orthonormalized and have the time dependence of the QS type. Indeed, $f(t) = f(t + T)$ and consequently

$$\int_{t_0}^t f(\tau)\,d\tau = \text{const} + f_0 t + g(t), \qquad f_0 \equiv \frac{1}{T}\int_0^T f(\tau)\,d\tau, \qquad g(t) = g(t + T).$$

$$\tag{2.3.6}$$

Note the connection between the Fourier coefficients

$$g_N = \mathrm{i} f_N/(\omega N), \quad f(t) = \sum_N f_N \exp(-\mathrm{i}N\omega t),$$

$$g(t) = \sum_{N \neq 0} g_N \exp(-\mathrm{i}N\omega t). \tag{2.3.7}$$

Substituting (2.3.6) into (2.3.4) we obtain $\psi_\alpha(x, t)$ in the QS form:

$$\psi_\alpha(x, t) = \exp(-\mathrm{i}\varepsilon_\alpha t) \cdot \varphi_\alpha(x, t),$$

$$\varepsilon_\alpha = \lambda U_\alpha f_0, \qquad \varphi_\alpha(x, t) = |\alpha\rangle \, \exp[-\mathrm{i}\lambda U_\alpha g(t)]. \tag{2.3.8}$$

From (2.3.5) and (2.3.8) it is easy to see that

$$(\lambda V - \mathrm{i}\partial/\partial t)\varphi_\alpha(x, t) = \varepsilon_\alpha \varphi_\alpha(x, t). \tag{2.3.9}$$

Therefore the set of eigen-QS of the operator $\lambda V - \mathrm{i}\,\partial/\partial t$, in the Hilbert space $L_2(x \otimes t)$ of time-periodic functions $\varphi(x, t)$ with the scalar product $\langle\langle \varphi_1 | \varphi_2 \rangle\rangle$ of (2.2.3), is

$$|N\alpha\rangle\rangle = |\alpha\rangle \, \exp[-\mathrm{i}\lambda U_\alpha g(t) + \mathrm{i}N\omega t]$$

$$(N = 0, \pm1, \pm2, \ldots),$$

$$(\lambda V - \mathrm{i}\,\partial/\partial t)|N\alpha\rangle\rangle = (\lambda U_\alpha f_0 + N\omega)|N_\alpha\rangle\rangle. \tag{2.3.10}$$

We shall try to find the solutions of the Schrödinger equation $\mathrm{i}\,\partial\psi/\partial t = (H + \lambda V)\psi$ in the QS form

$$\tilde{\psi}_p(x, t) = \exp(-\mathrm{i}\tilde{\varepsilon}_p t)\tilde{\varphi}_p(x, t), \quad (H - \lambda V - \mathrm{i}\,\partial/\partial t)\tilde{\varphi}_p = \tilde{\varepsilon}_p \tilde{\varphi}_p. \tag{2.3.11}$$

We seek $\tilde{\varphi}_p(x, t)$ in the form

$$\tilde{\varphi}_p(x, t) = \sum_{N,\alpha} C_p^{N\alpha} |N\alpha\rangle\rangle. \tag{2.3.12}$$

Substituting (2.3.12) into the second line of (2.3.11), allowing for (2.3.10) and taking the scalar product with some arbitrary QS $\langle\langle M\beta|$ from (2.3.10), we obtain the usual set of linear equations for determining $C_p^{N\alpha}$ and $\tilde{\varepsilon}_p$ (at arbitrary M, β):

$$\sum_{N,\alpha} \left\{ C_p^{N\alpha}[\langle\langle M\beta|H|N\alpha\rangle\rangle + \delta_{MN}\delta_{\alpha\beta}(\lambda U_\alpha f_0 + N\omega)] \right\} = \tilde{\varepsilon}_p C_p^{M\beta}. \tag{2.3.13}$$

The matrix elements $\langle\langle M\beta|H|N\alpha\rangle\rangle$ may be calculated more explicitly by substituting (2.3.10) into (2.2.3):

$$\langle\langle M\beta|H|N\alpha\rangle\rangle = \langle\beta|H|\alpha\rangle \frac{1}{T} \int\limits_0^T \exp[\mathrm{i}(N - M)\omega t + \mathrm{i}\lambda(U_\beta - U_\alpha)g(t)]\,\mathrm{d}t. \tag{2.3.14}$$

From (2.3.13) it is seen that the *sought quasienergies* $\tilde{\varepsilon}_p$ *and the expansion coefficients* $C_p^{N\alpha}$ *represent, respectively, the eigenvalues and the eigenvectors of* the matrix $\tilde{H}_{M\beta}^{N\alpha}$ which is constructed from the QSs $|N\alpha\rangle$ (2.3.10):

$$\tilde{H}_{M\beta}^{N\alpha} \equiv \langle\beta|H|\alpha\rangle\frac{1}{T}\int\limits_0^T dt \, \exp[i(N-M)\omega t + i\lambda(U_\beta - U_\alpha)g(t)]$$

$$+ \delta_{MN}\delta_{\alpha\beta}(\lambda U_\alpha f_0 + N\omega). \tag{2.3.15}$$

At $\lambda \to \infty$ (very strong field) the diagonal elements of the $\tilde{H}_{M\beta}^{N\alpha}$ matrix (2.3.15) are either constant (at $f_0 U_\alpha = 0$) or increasing (at $f_0 U_\alpha \neq 0$). The non-diagonal elements decrease since the greater λ is the more rapidly the integrand in (2.3.15) oscillates.

As an example consider the function $\lambda V(x,t)$ of the type $\lambda V(x,t) = \lambda U(x)\cos\omega t$. For $f(t) = \cos\omega t$ we find

$$f_0 = 0, \qquad g(t) = \omega^{-1}\sin\omega t \tag{2.3.16}$$

and time integration in (2.3.15) gives

$$\frac{1}{T}\int\limits_0^T dt \, \exp[i(N-M)\omega t$$

$$+ i\lambda(U_\beta - U_\alpha)\omega^{-1}\sin\omega t] = J_{M-N}(\lambda\omega^{-1}(U_\beta - U_\alpha)), \tag{2.3.17}$$

where $J_k(y)$ is a Bessel function. So to find the QSs the following matrix must be diagonalized:

$$\tilde{H}_{M\beta}^{N\alpha} \equiv \langle\beta|H|\alpha\rangle J_{M-N}(\lambda\omega^{-1}(U_\beta - U_\alpha)) + \delta_{MN}\delta_{\alpha\beta}N\omega. \tag{2.3.18}$$

2.3.2 Perturbation Theory

Note, first of all, that the PT of QSs for the $\tilde{H}_{M\beta}^{N\alpha}$-matrix depends on the value of $f_0 \equiv T^{-1}\int_0^T f(\tau)\,d\tau$. If $f_0 = 0$, then the diagonal elements $\lambda U_\alpha f_0 + N\omega$ at different α in general differ, i.e., the matrix in (2.3.15) is nondegenerate. If $f_0 = 0$, then the diagonal elements are equal to $N\omega$ and do not depend on α, i.e., the matrix in (2.3.15) is degenerate.

We shall construct the PT for the most interesting case (for real applications) of $f_0 = 0$. The "unperturbed" (i.e. at $H = 0$) QSs $\varphi_\alpha(x,t)$ are degenerate: $\varepsilon_\alpha = 0$. Therefore it is necessary first to find CSZO representing some linear combinations of φ_α and then to calculate the corrections to quasienergies and to φ_p.

We introduce a parameter h into the nondiagonal elements (2.3.14) through the substitution $H \to hH$. Then the $\tilde{H}_{M\beta}^{N\alpha}$ matrix (2.3.15) becomes

$$\tilde{H}_{M\beta}^{N\alpha} = h\langle\beta|H|\alpha\rangle\frac{1}{T}\int\limits_0^T dt \, \exp[i(N-M)\omega t + i\lambda(U_\beta - U_\alpha)g(t)]$$

$$+ \delta_{\alpha\beta}\delta_{MN}N\omega. \tag{2.3.19}$$

We try to find $\tilde{\varepsilon}_p$ and $\tilde{\varphi}_p$ (2.3.11) in the form of the series in powers of h (in the zeroth order $\tilde{\varepsilon}_p = \varepsilon_\alpha = 0$):

$$\varepsilon_p = h\varepsilon_p^{(1)} + h^2\varepsilon_p^{(2)} + \cdots, \qquad \tilde{\varphi}_p = \sum_{N,\alpha} C_p^{N\alpha}|N\alpha\rangle;$$

$$C_p^{N\alpha} = \delta_{N_0}C_p^\alpha + hA_p^{N\alpha} + h^2B_p^{N\alpha} + \cdots. \tag{2.3.20}$$

In the zeroth order of $C_p^{N\alpha}$, we set $C_p^{N\alpha} = \delta_{N_0}C_p^\alpha$ in order to have as CSZO φ_p the linear combinations of QSs $|N\alpha\rangle$ (2.3.10) with $N = 0$, i.e., we want the QSs φ_α of (2.3.8). Substituting (2.3.20) into (2.3.13) and writing out the terms of the same order of h, we obtain the PT equations.

The zeroth order: for all M, β we have $C_p^\beta \delta_{M_0} M\omega = 0$.

The first order:

$$\sum_\alpha C_p^\alpha \langle\langle M\beta|H|0\alpha\rangle\rangle + A_p^{M\beta} M\omega = \varepsilon_p^{(1)}\delta_{M_0}C_p^\beta \tag{2.3.21}$$

for all M, β. Substituting $M = 0$ in (2.3.21) we obtain the *secular equation* for determining the CSZO φ_p

$$\varepsilon_p^{(1)}C_p^\beta = \sum_\alpha C_p^\alpha \langle\langle 0\beta|H|0\alpha\rangle\rangle. \tag{2.3.22}$$

We introduce a matrix $\tilde{H}_\alpha^\beta \equiv \langle\langle 0\beta|H|0\alpha\rangle\rangle$ constructed from the QSs φ_α [which diagonalize the operator $U(x)$]:

$$\tilde{H}_\alpha^\beta \equiv \langle\langle 0\beta|H|0\alpha\rangle\rangle = \langle\beta|H|\alpha\rangle \frac{1}{T} \int_0^T \exp[i\lambda(U_\beta - U_\alpha)g(t)]\,dt \tag{2.3.23}$$

From (2.3.22) it can be seen that the eigenvalues of the $\tilde{H}_\alpha^\beta$ matrix determine the energy corrections $\varepsilon_p^{(1)}$ and the eigenvectors of this matrix determine the coefficients C_p^α of the expansion of the CSZO φ_p on the QSs φ_α (in the first order of h).

Substituting $M \neq 0$ in (2.3.21) we obtain the formula for the first order of expansion coefficients

$$A_p^{M\beta} = -(M\omega)^{-1}\sum_\alpha C_p^\alpha \langle\langle M\beta|H|0\alpha\rangle\rangle. \tag{2.3.24}$$

Note that if the CSZO φ_p are found, all higher order corrections of PT may be determined by the formulas of the usual stationary PT, in which it is necessary to substitute $N\omega$ instead of unperturbed energies and to use the matrix elements $\langle\langle Np|H|Mq\rangle\rangle$ calculated according to (2.2.3).

Consider as an example the important case of $V(x,t) = U(x)\cos\omega t$. Substituting $g(t) = \omega^{-1}\sin\omega t$ into (2.3.23) and integrating we obtain

$$\tilde{H}_\alpha^\beta = \langle\beta|H|\alpha\rangle J_0(\lambda\omega^{-1}(U_\beta - U_\alpha)). \tag{2.3.25}$$

Note that the analogous matrix was obtained in [2.8] by the so-called "averaging principle". We see that in our general approach the matrix of type (2.3.25) arises only as the first-order correction [and only for the particular case $V(x, t) = U(x) \cos \omega t$].

The criterion of the applicability of PT is, as usual, the smallness of the quasienergy corrections $\varepsilon_p^{(1)}$ in comparison to the distance between unperturbed (at $H = 0$) levels, i.e. in our case in comparison to the field frequency ω.

Thus, the QSs of a quantum system with the Hamiltonian $H(x)+\lambda U(x) f(t)$ of (2.3.1) may be found in the limit $\lambda \to \infty$ by the following scheme.

1. The matrix of the operator $U(x)$ is constructed from the eigenstates $|n\rangle$ of the Hamiltonian H and is diagonalized. Its eigenvalues U_α and eigenvectors C_n^α determine the states $|\alpha\rangle$ (2.3.3) (the linear combinations of the states $|n\rangle$) in which the operator $U(x)$ is diagonal.

2. The matrix $\tilde{H}_\alpha^\beta$ is calculated by (2.3.23) in which the function $g(t)$ is defined by (2.3.6).

3. The matrix $\tilde{H}_\alpha^\beta$ is diagonalized. Its eigenvalues $\varepsilon_p^{(1)}$ give the quasienergies (in the limit $\lambda \to \infty$) and its eigenvectors C_p^α determine the expansion coefficients of the corresponding QS $\tilde{\varphi}_p(x, t)$ on the states $\varphi_\alpha(x, t)$ (2.3.8):

$$\tilde{\varphi}_p(x, t) = \sum_\alpha C_p^\alpha \varphi_\alpha(x, t). \qquad (2.3.26)$$

2.3.3 Generalizations

Consider the Hamiltonian of a more general form

$$\tilde{H}(x, t) = H(x) + W(x, t) + \lambda U(x) f(t),$$
$$f(t + T) = f(t), \qquad W(x, t + T) = W(x, t), \qquad (2.3.27)$$

where $W(x, t)$ is some additional time-periodic perturbation. The relation between all the values in question is assumed to be

$$\max(\omega, (\lambda V \omega)^{1/2}) >> \Delta \sim W. \qquad (2.3.28)$$

According to (2.3.28) the additional perturbation W is not, in general, small compared to the characteristic level separation Δ.

All the results of Sects. 2.3.1, 2 are immediately applicable to this case: QSs for (2.3.27) are sought in the form of the expansion (2.3.12) on the set of QSs in the field corresponding to the term $V(x, t)$ only, see (2.3.10). The quasienergies $\tilde{\varepsilon}_p$ and the expansion coefficients $C_p^{N\alpha}$ are determined by the eigenvalues and the eigenvectors of the matrix

$$\tilde{H}_{M\beta}^{N\alpha} = \langle\langle M\beta|H + W(t)|N\alpha\rangle\rangle + \delta_{MN}\delta_{\alpha\beta}(\lambda U_\alpha f_0 + N\omega). \qquad (2.3.29)$$

In particular, for the stationary perturbation $W(x)$ the matrix (2.3.28) is reduced to

$$\tilde{H}_{M\beta}^{N\alpha} = \langle\beta|H + W|\alpha\rangle \frac{1}{T} \int_0^T dt \, \exp[i(N - M)\omega t + i\lambda(U_\beta - U_\alpha)g(t)]$$
$$+ \delta_{MN}\delta_{\alpha\beta}(\lambda U_\alpha f_0 + N\omega). \qquad (2.3.30)$$

The PT of Sect. 2.3.2 is generalized analogously: For the case $f_0 = 0$ QS are sought in the form (2.3.20), where the first-order corrections $\varepsilon_p^{(1)}$ to the quasienergies and the expansion coefficients of the CSZO are determined by the matrix

$$\tilde{H}_\alpha^\beta = \langle\langle 0\beta|H + W(t)|0\alpha\rangle\rangle. \tag{2.3.31}$$

The criterion of PT applicability is the smallness of the corrections $\varepsilon_p^{(1)}$ relative to the frequency ω. It can be shown that the sufficient condition of the smallness of the corrections can be represented in the form of (2.3.28).

3 Action of One-Dimensional Quasimonochromatic Electric Fields on Coulomb Emitters

The first theoretical paper on the calculation of the radiation spectrum of a quantum system under the action of QEFs was published in 1933 by *Blochinzew* [3.1]. He analyzed the splitting of a model hydrogen line (consisting of only one Stark component) in a field $E_0 \cos \omega t$ and showed that the line splits into, in general, an infinite number of satellites separated from the line center by the frequencies $\Delta \omega = \pm \omega, \pm 2\omega, \ldots, \pm p\omega, \ldots$. Actually QSs of a quantum system were found for the first time in [3.1] (without using this terminology). In this chapter, further results concerning the profiles of hydrogen-like spectral line under the action of a single- or multimode one-dimensional QEF are given. The possibilities for the implementation of these results in plasma diagnostics are also discussed. The main results of Chap. 3 are contained in [3.2–9].

3.1 Splitting of Hydrogen-like Spectral Lines in a Single-Mode QEF

3.1.1 Analytical Investigation

Model (one component) hydrogen SL. In a "reduced frequency" scale the SL profile $S_B(\Delta \omega / \omega)$ obtained in [3.1] may be expressed in the form (Appendix A)

$$S_B \left(\frac{\Delta \omega}{\omega} \right) = \sum_{p=-\infty}^{+\infty} \delta \left(\frac{\Delta \omega}{\omega} - p \right) J_p^2(X\varepsilon), \qquad \int_{-\infty}^{+\infty} d \left(\frac{\Delta \omega}{\omega} \right) S_B \left(\frac{\Delta \omega}{\omega} \right) = 1,$$

$$\varepsilon \equiv 3\hbar E_0 / (2m_e e\omega), \qquad X \equiv n(n_1 - n_2) - n'(n_1' - n_2'), \qquad (3.1.1)$$

where $J_p(z)$ is a Bessel function. Of most interest to practical applications is the case of strong phase modulation $(X\varepsilon \gg 1)$ in which the form of the envelope of the satellites is significant.

We find the form of the envelope of the satellites by using the asymptotic relation of Bessel functions of large arguments and indices to the Airy function [3.10]

$$J_\nu(\nu + z\nu^{1/3}) \approx (2/\nu)^{1/3} \mathrm{Ai}(-2^{1/3}z), \qquad \nu \gg 1. \qquad (3.1.2)$$

After replacing $p = v$, $y = v + zv^{1/3}$ it is easy to find

$$\frac{\partial J_p(y)}{\partial p} \approx -\frac{\partial J_p(y)}{\partial y} \approx (2/v)^{2/3} \text{Ai}'(-2^{1/3}z), \quad v \gg 1. \tag{3.1.3}$$

From (3.1.3) it is seen that the stationary points of the function $J_p(y)$ of two variables are approximately determined by the equation $\text{Ai}'(-2^{1/3}z) = 0$, from which we find

$$z = -2^{-1/3}a'_s, \qquad y = p - a'_s(p/2)^{1/3}, \tag{3.1.4}$$

where the a'_s are zeros of the tabulated Airy function derivative ($s = 1, 2, 3, \ldots$) [3.10]. Thus at a fixed E_0/ω ratio (and correspondingly at fixed $y = 3\hbar X E_0/2m_e e\omega$) the envelope of the satellites $I(p) = J_p^2(y)$ is an oscillatory function of p. The positions of the envelope maxima p_s and the maxima values $I_{\max}$ are determined by

$$p_s \approx y + a'_s(y/2)^{1/3}, \qquad I_{\max}^{(S)} \approx [\text{Ai}(a'_s)]^2 (2/y)^{2/3}. \tag{3.1.5}$$

The numerical coefficients in (3.1.5) $2^{-1/3}a'_s$, $2^{2/3}[\text{Ai}(a'_s)]^2$ are equal to -0.809 and 0.4555, respectively, for the first maximum and -2.58 and 0.2787 for the second maximum.

Note that the first maximum is 1.6 times higher than the second one and they are separated by $\Delta\omega_{\max}^{I-II} \approx 1.8y^{1/3}\omega$. Thus, with increasing field amplitude, more and more satellites concentrate inside each peak, so that the oscillations of the envelope become more and more pronounced. As an example, in Fig. 3.1 the exact SL profile in the vicinity of the first maximum and the approximate envelope [according to (3.1.2)] are shown for $y = 27$. The total number of envelope maxima (in each line wing) is determined from $\tilde{s} \approx 0.300y + 0.75$, which demonstrates very good accuracy not only for $y \gg 1$ but even at $y \sim 1$.

The same problem may be considered in the quasistatic (q.s.) limit ($\omega \to 0$), in which the profile of a one component hydrogen SL $S_{\text{q.s.}}(\Delta\omega/\omega)$ repeats the

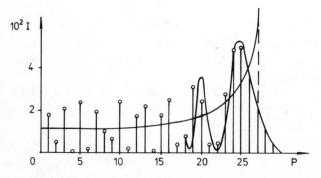

Fig. 3.1. Comparison of the approximate Airy-type envelope of satellites (oscillating curve) with the exact profile of a Stark component at $X\varepsilon = 27$. The monotonic curve represents the static profile

distribution of the instantaneous field intensity:

$$S_{\text{q.s.}}(\Delta\omega/\omega) = (\omega/\pi)[(CE_0)^2 - (\Delta\omega)^2]^{-1/2}, \quad C \equiv 3X_{\alpha\beta}\hbar/(2m_e e).$$
$$(3.1.6)$$

The following question may arise: Why, even at strong modulation ($y = CE_0/\omega \gg 1$) does the envelope of the Blochinzew profile $S_B(\Delta\omega/\omega)$ oscillate so strongly and differ from the quasistatic profile $S_{\text{q.s.}}(\Delta\omega/\omega)$?

To answer this question we analyze the validity conditions of the profile (3.1.6) following *Schrödinger* [3.11]. In the adiabatic approximation the dipole moment matrix element $d_{\alpha\beta}(t)$ between the initial and final Stark states of an atom contains the factor $\exp[iC \int_0^t E(t)\,dt]$. In the low-frequency limit ($\omega \to 0$) the field $E(t)$ may be represented as $E_0 \cos\omega t \approx E_0 - E_0\omega^2 t^2/2$. In this approximation we have

$$d_{\alpha\beta}(t) \propto \exp(iCE_0 t - iCE_0\omega^2 t^3/6). \tag{3.1.7}$$

In the case of

$$t \ll (CE_0\omega^2)^{-1/3} \equiv t_{\text{cr}} \tag{3.1.8}$$

the second term in the exponent (3.1.7) is small compared to unity and the first term corresponds to the static result.

Consequently the quasistatic profile (3.1.6) is valid for small times $t \ll t_{\text{cr}}$. Physically this means that t_{cr} is large compared to the observation time t_{exp} or to the atomic lifetime in an excited state t_{life}. We designate $\tau^{-1} = t_{\text{exp}}^{-1} + t_{\text{life}}^{-1}$. Then the quasistatic profile (3.1.6) corresponds to the case when $(CE_0\omega^2)^{1/3}\tau \ll 1$ and the profile (3.1.1) corresponds to the case when $(CE_0\omega^2)^{1/3}\tau \gg 1$ (at $CE_0 \gg \omega$).

Thus the validity of the Blochinzew profile is in principle restricted at low frequencies by the condition $\omega \gg (CE_0\tau^3)^{-1/2}$. Therefore it is no wonder that if we let $\omega \to 0$ the Blochinzew profile does not transform into the quasistatic profile (3.1.6).

Real (multicomponent) hydrogen SL. The line profile may be represented

$$S(\Delta\omega/\omega) = \sum_{p=-\infty}^{+\infty} I(p, \varepsilon)\delta(\Delta\omega/\omega - p); \quad \int_{-\infty}^{+\infty} d(\Delta\omega/\omega)S(\Delta\omega/\omega) = 1;$$

$$I(p, \varepsilon) = \left(f_0 + 2\sum_{k=1}^{k_m} f_k\right)^{-1} \left[f_0\delta_{p0} + 2\sum_{k=1}^{k_m} f_k J_p^2(X_k\varepsilon)\right] \tag{3.1.9}$$

(f_0 is the total intensity of all central Stark components, f_k is the intensity of the lateral component with the number $k = 1, 2, \ldots, k_m$).

In the case of weak modulation ($\langle X_k\rangle\varepsilon \ll 1$), up to terms of the order of ε^2 inclusive we have

$$S(\Delta\omega/\omega) \approx I(0, \varepsilon)\delta(\Delta\omega/\omega) + I(\pm 1, \varepsilon)[\delta(\Delta\omega/\omega - 1) + \delta(\Delta\omega/\omega + 1)],$$
$$I(0, \varepsilon) = 1 - 2A\varepsilon^2, \quad I(\pm 1, \varepsilon) = A\varepsilon^2; \quad A \equiv \langle X_k^2\rangle/4. \tag{3.1.10}$$

For the first four Balmer SLs the constants A in (3.1.10) are 1.42089, 12.5215, 40.6098, 96.4440 for H_α, H_β, H_γ, and H_δ respectively.

The case of most practical interest is that of strong modulation ($\langle X_k \rangle \varepsilon \gg 1$), in which the form of the envelope of the satellites is important. From (3.1.5) it follows that the absolute maximum of the envelope for the Stark component with the number k corresponds to the satellite with the number

$$p_1(X_k\varepsilon) \approx X_k\varepsilon - 0.809(X_k\varepsilon)^{1/3} \tag{3.1.11}$$

and the value of the maximum is $I_{\max}^{(1)} \approx 0.4555(X_k\varepsilon)^{-2/3}$. Therefore, the maxima and the halfwidth of the whole profile $S(\Delta\omega/\omega)$ are determined by the maxima and the halfwidth of the envelope of the following *effective* profile:

$$S_{\mathrm{eff}}(\Delta\omega/\omega) = \left(f_0 + 2\sum_{k=1}^{k_m} f_k \right)^{-1} \left\{ f_0\delta(\Delta\omega/\omega) + \sum_{k=1}^{k_m} 0.911 f_k (X_k\varepsilon)^{2/3} \right.$$

$$\left. [\delta(\Delta\omega/\omega - p_1(X_k\varepsilon)) + \delta(\Delta\omega/\omega + p_1(X_k\varepsilon))] \right\}. \tag{3.1.12}$$

Two important corollaries follow from (3.1.12).

Corollary 1. *In the case* $\langle X_k \rangle \varepsilon \gg 1$, *for the SLs with central components* ($f_0 \neq 0$, *e.g. for* H_α, H_γ, H_ε), *the ratio of the intensities of the central peak to the lateral one increases proportionally to* $\varepsilon^{2/3}$ *with increasing* ε. *Hence the envelope halfhalfwidth* $p_{1/2}$ *tends to its minimal value* $p_{1/2}^{\min} = 1/2$. *The presence of the lateral components may be manifested only in the far wings at intensities* $I/I_{\max} \lesssim (\langle X_k \rangle \varepsilon)^{-2/3}$.

Corollary 2. *In the case* $\langle X_k \rangle \varepsilon \gg 1$ *for the SLs without central components* (*e.g.* H_β, H_δ, H_ζ) *the effective profile reproduces the picture of the static Stark splitting in the field* E_0 *but with redefined component intensities:*

$$f_k^{\mathrm{eff}} \equiv f_k X_k^{-2/3}. \tag{3.1.13}$$

This leads, generally speaking, to drastic changes in the positions of the maxima and the envelope halfwidth of the whole SL compared with the static Stark profile (Fig. 3.2).

Corollary 2 allows us to obtain, for any hydrogen SL, simple analytical expressions for the position of the satellite $p_{\max} = \Delta\omega_{\max}/\omega$ having the maximum intensity and for the envelope halfhalfwidth $p_{1/2} = \Delta\omega_{1/2}/\omega$. [The value $p_{1/2}$ corresponds to the position of the component with the number $\tilde{k}$ which is determined from the conditions $f_{\tilde{k}}^{\mathrm{eff}} \geq f_{\max}^{\mathrm{eff}}/2$, $f_{\tilde{k}+i}^{\mathrm{eff}} < f_{\max}^{\mathrm{eff}}/2$ ($i = 1, 2, \ldots, k_m - \tilde{k}$).] The following analytical results correspond to the case of transverse observation with respect to the vector E:

$$p_{\max}(H_\beta) \approx 4\varepsilon - 1.28\varepsilon^{1/3}, \qquad p_{1/2}(H_\beta) \approx 8\varepsilon;$$

$$p_{\max}(H_\delta) \approx 6\varepsilon - 1.47\varepsilon^{1/3}, \qquad p_{1/2}(H_\delta) \approx 6\varepsilon;$$

$$p_{\max}(H_\zeta) \approx 8\varepsilon - 1.62\varepsilon^{1/3}, \qquad p_{1/2}(H_\zeta) \approx 8\varepsilon. \tag{3.1.14}$$

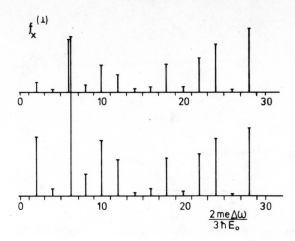

Fig. 3.2. Comparison of the profile S_{stat} of the H_δ line in a static field E_0 (*upper graph*) and the effective profile S_{eff} of the H_δ line in a field $E_0 \cos \omega t$ (*lower graph*)

We write out the analogous parameters for the static profiles of the same SL:

$$p_{max}^{stat}(H_\beta) = 4\varepsilon, \qquad p_{1/2}^{stat}(H_\beta) \approx 10\varepsilon;$$

$$p_{max}^{stat}(H_\delta) \approx p_{1/2}^{stat}(H_\delta) \approx 28\varepsilon; \qquad p_{max}^{stat}(H_\zeta) \approx p_{1/2}^{stat}(H_\zeta) \approx 54\varepsilon. \quad (3.1.15)$$

Thus for the case of $\langle X_k \rangle \varepsilon \gg 1$ for the lines H_δ, H_ζ the positions of the maxima and the profile halfwidth in the field $E_0 \cos \omega t$ substantially differ (5–7 times) from the corresponding values for the profiles in the static field E_0 (H_δ profiles in Fig. 3.2).

The effective profile of hydrogen-like SL for ions of charge Z in a laser field $E_0 \cos \omega t$ is described by (3.1.11, 12) with $X_k \varepsilon / Z$ replacing $X_k \varepsilon$. For the lines L_β and L_δ we obtain (at $\varepsilon / Z \gtrsim 1$):

$$p_{max}(L_\beta) \approx 3\varepsilon/Z - 1.17(\varepsilon/Z)^{1/3}, \qquad p_{1/2}(L_\beta) \approx 6\varepsilon/Z;$$

$$p_{max}(L_\delta) \approx 5\varepsilon/Z - 1.38(\varepsilon/Z)^{1/3}, \qquad p_{1/2}(L_\delta) \approx 20\varepsilon/Z. \quad (3.1.16)$$

3.1.2 Numerical Calculations. Oscillatory Behavior of Halfwidths and Intensities

We have numerically calculated the Blochinzew profiles (3.1.9) of the main Balmer SLs H_n ($n = 3, \ldots, 8$) in the observation direction transverse to E_0. The step on the ε scale was chosen rather small: $\delta\varepsilon = 0.0125$. The results are reported in [3.4] (some of them are presented in Appendix A). For each profile the parameter $p_{1/2} = \Delta\omega_{1/2}^{half}/\omega$ which is the halfhalfwidth of the profile arising after the connection of each two neighbor satellites by straight line segments was also calculated.

The results of the numerical calculations of $p_{1/2}(\varepsilon)$ are shown in Figs. 3.3–6. From Fig. 3.3 it is seen that the calculation confirms Corollary 1: for the lines H_α, H_γ, H_ε at $\varepsilon > 0.65$ the halfhalfwidth $p_{1/2}(\varepsilon)$ decreases (oscillating a little) with increase of ε and tends to $p_{1/2}^{min} = 0.5$.

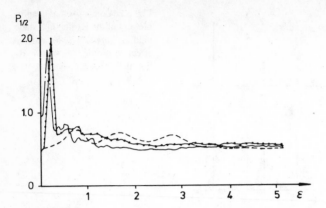

Fig. 3.3. Dependence of halfhalfwidths $p_{1/2}$ of the envelopes for numerically calculated profiles of H_α (*dashed line*), H_ε (*solid line*), H_γ (*solid line with dots*) in a field $E_0 \cos \omega t$ upon the parameter $\varepsilon = 3\hbar E_0 / 2m_e e\omega$

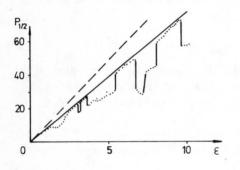

Fig. 3.4. The same as in Fig. 3.3 but for H_β (*dots*). The analytical dependences $p_{1/2}(\varepsilon)$ for the profile S_{eff} in a field $E_0 \cos \omega t$ (*solid line*) and for the profile S_{stat} in a field E_0 (*dashed line*) are also displayed

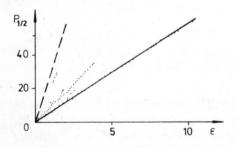

Fig. 3.5. The same as in Fig. 3.4 but for H_δ

It is interesting to compare the analytical and numerical results at $\varepsilon \gtrsim 1$ for the lines containing no central components. From Figs. 3.4–6 it is easy to conclude that the analytical dependences $p_{1/2}(\varepsilon)$ given by (3.1.14) [but not the $p_{1/2}^{\text{stat}}(\varepsilon)$ of (3.1.15)] agree well with the numerical results. An analogous conclusion can be drawn from the comparison of the absolute maxima of the profiles. These results confirm Corollary 2.

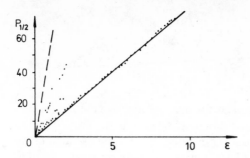

Fig. 3.6. The same as in Fig. 3.4 but for H_ζ

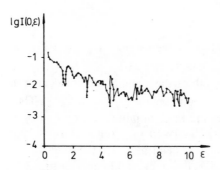

Fig. 3.7. Dependence of the intensity at the center of numerically calculated profiles of H_β upon the parameter ε

From Figs. 3.3–6 it is seen that the halfhalfwidths $p_{1/2}$ are not monotonic functions of the field amplitude E_0, but experience numerous oscillations. This is not connected with the chosen definition of $p_{1/2}$ since the envelope maximum $\max_p I(p, \varepsilon)$ also turns out to be a nonmonotonic function of E_0.

If we fix the separation from the SL center (i.e. tune to a certain wavelength λ) and change E_0, then the light intensity $I_\lambda(E_0)$ will also be a nonmonotonic oscillatory function of E_0. In Fig. 3.7 such a dependence for the intensity in the center of the H_β line ($\lambda = \lambda_0 = 486.13$ nm) is shown. It can be seen that near some values of $\varepsilon = \varepsilon_{\min}$ a sharp intensity change (of one order of magnitude) occurs when E_0 changes by only a few percent. These oscillations are the consequence of the oscillatory dependence of the Bessel functions $J_p(X_k\varepsilon)$ in (3.1.9) upon their arguments.

The effect of intensity oscillations in the center of the lines H_β, H_δ, H_ζ can be of practical use in nonlinear optics. A system "hydrogen atoms + microwave field" inserted into an optical resonator can be a multifunctional optical element with tunable parameters. In particular, due to nonmonotonic kangaroo-type dependence of I_{λ_0} on E_0 in such a system, interesting threshold phenomena may be observed.

3.1.3 Doppler Broadening. Formal Analogy with Thomson Scattering in the Presence of QEFs

The profile of a Stark component j of a hydrogen SL in a field $E_0 \cos \omega t$ with allowance for Doppler broadening has the form

$$I_j(\Delta\omega) = \sum_{p=-\infty}^{+\infty} J_p^2(y_j)\frac{1}{\pi^{1/2}\Delta\omega_T}\exp\left[-\left(\frac{\Delta\omega - p\omega}{\Delta\omega_T}\right)^2\right],$$

$$y_j \equiv X_j\varepsilon, \qquad \Delta\omega_T \equiv \frac{\omega_0}{c}\left(\frac{2T_a}{M_a}\right)^{1/2}, \tag{3.1.17}$$

where T_a and M_a are the temperature and the mass of hydrogen atoms.

First consider the "dynamic" case where $(\langle X_j\rangle\varepsilon)^{1/3}\omega \gg \Delta\omega_T \gtrsim \omega$, so that the thermal motion does not smooth over satellite envelope oscillations. In this case the intensity of the profile (3.1.17) at the absolute maxima located at frequencies $\Delta\omega \approx \pm y_j\omega$ is

$$I_j(\pm y_j\omega) \approx 0.911\omega^{-1}y_j^{-2/3}. \tag{3.1.18}$$

Hence, for the SL containing no central components (e.g. H$_\beta$, H$_\delta$, H$_\zeta$) the maxima and the halfwidth of the whole profile $I(\Delta\omega) = 2\sum_{j=1}^{j_{\max}} f_j I_j(\Delta\omega)$ are determined by the maxima and the halfwidth of the static Stark broadening in the field E_0 but with redefined intensities of components: $f_j^{\text{eff}} = f_j X_j^{-2/3}$. Thus it is possible to determine experimentally the field amplitude E_0 by hydrogen SL halfwidths using the relation $\Delta\omega_{1/2} = p_{1/2}\omega$ [with $p_{1/2}$ from (3.1.14) so that $\Delta\omega_{1/2} \propto \varepsilon\omega$ does not depend on ω] and then to determine also the frequency ω from lateral maxima widths $\Delta\omega_{1/2}^{\max} \sim (\langle X_j\rangle\varepsilon)^{1/3}\omega$ (cf. solid curve in Fig. 3.8)

Now consider the case where $(\langle X_j\rangle\varepsilon)^{1/3}\omega \ll \Delta\omega_T \ll \varepsilon\langle X_j\rangle\omega$. In this case the thermal motion, in fact, covers up the dynamic character of the modulating field. The profile of the component may be represented as

$$I_j(\Delta\omega) = \int_0^{2\pi}(d\Phi/2\pi)\pi^{-1/2}(\Delta\omega_T)^{-1}\exp[-(\Delta\omega - y_j\omega\sin\Phi)^2/(\Delta\omega_T)^2].$$

$$\tag{3.1.19}$$

In the maxima (of which there are only two now) symmetrically located at the frequencies $\Delta\omega \approx \pm y_j\omega$ we find

$$I_j(\pm y_j\omega) \approx \Gamma(1/4)\pi^{-3/2}(2\Delta\omega_T y_j\omega)^{-1/2}, \quad \Gamma(1/4) \approx 3.626. \tag{3.1.20}$$

Hence, instead of $f_j^{\text{eff}} = f_j X_j^{-2/3}$ the values $\tilde{f}_j^{\text{eff}} = f_j X_j^{-1/2}$ should be taken. The results for SL halfhalfwidths are

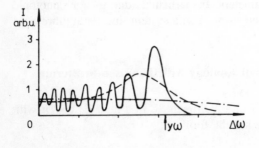

Fig. 3.8. The profiles of a hydrogen line Stark component (or a Thomson scattering line) in a field $E_0\cos\omega t$ with allowance for Doppler broadening: *solid line*, $(\langle X_j\rangle\varepsilon)^{1/3}\omega \gg \Delta\omega_T \gtrsim \omega$; *dashed line*, $(\langle X_j\rangle\varepsilon)^{1/3}\omega \ll \Delta\omega_T \ll \varepsilon\langle X_j\rangle\omega$; *dashed-dotted line*, $\langle X_j\rangle\varepsilon\omega \ll \Delta\omega_T$

$$\Delta\omega_{1/2}(H_\beta) \approx 10\varepsilon\omega, \quad \Delta\omega_{1/2}(H_\delta) \approx 28\varepsilon\omega, \quad \Delta\omega_{1/2}(H_\zeta) \approx 54\varepsilon\omega.$$

$$(3.1.21)$$

Maxima halfwidths are approximately equal to $(2\omega_0/c)[(2 \ln 2)T_a/M_a]^{1/2}$. Thus in this case both the amplitude E_0 and the temperature T_a could be experimentally determined (dashed curve in Fig. 3.8).

Finally, in the third case, where $\langle X_j \rangle \varepsilon\omega \ll \Delta\omega_T$, the profile (3.1.17) is well described by a Gauss distribution (dashed-dotted curve in Fig. 3.8). Thus it is hard to obtain information on field parameters; in principle it might be obtained by introducing a polarizer into the detection system.

Results of similar form but of a Doppler (not Stark) nature appear in at least two other types of experiments. In gamma-resonant spectroscopy, in the presence of an ultrasound field at a frequency ω, owing to the Doppler effect, the γ-radiation is frequency modulated. In Thomson scattering (TS) of laser radiation at a frequency ω_0 from plasma electrons in the presence of a microwave field $E_0 \cos \omega t$, the light scattered by the electron oscillating in the field acquires a frequency modulation because of the Doppler effect [3.12]:

$$E_s \propto \cos\left(\omega_0 t + \frac{\delta\omega}{\omega}\sin\omega t\right), \quad \delta\omega \equiv 2\omega_0 \frac{V_{osc}}{c}\left(\sin\frac{\theta}{2}\right)\cos\varphi,$$

$$V_{osc} = \frac{eE_0}{m_e(\omega^2 + \nu^2)^{1/2}}, \tag{3.1.22}$$

where θ is the scattering angle, φ is the angle between the vectors E_0 and $(k_s - k_i)$ (k_s and k_i are wave vectors of the scattered and incident waves respectively), and ν is the collision frequency. The TS spectrum is described by (3.1.17) with $\delta\omega/\omega$ substituted for $X_j\varepsilon$ and $\Delta\omega_T = 2c^{-1}\omega_0 \sin(\theta/2)(2T_e/m_e)^{1/2}$. Assuming strong modulation $(\delta\omega/\omega) \equiv BE_0/\omega^2 \gg 1$ ($B \equiv 2e\omega_0)\times[\sin(\theta/2) \cos\varphi]/m_e c$, $\nu \ll \omega$) we have obtained the following results.

In the case $(BE_0\omega)^{1/3} \gg \Delta\omega_T \gtrsim \omega$, the line halfhalfwidth is $\Delta\omega_{1/2} \approx BE_0/\omega$ and the halfwidth of the maxima is $\Delta\omega_{1/2}^{max} \sim (BE_0\omega)^{1/3}$. Thus both the amplitude and the frequency could be experimentally determined.

However, in the case $(BE_0\omega)^{1/3} \ll \Delta\omega_T \ll BE_0/\omega$ we still have $\Delta\omega_{1/2} \approx BE_0/\omega$ but the halfwidth of the maxima is now $\Delta\omega_{1/2}^{max} \approx 2(\ln 2)^{1/2}\Delta\omega_T$. In this case the ratio E_0/ω and the temperature T_e could be experimentally determined.

3.2 Splitting of Hydrogen-like Spectral Lines in a Multimode QEF

3.2.1 Analytical Investigation for the Number of Modes Approaching Infinity, and Equal to 2

The multimode case. Consider the hydrogen SL splitting under the action of a one-dimensional QEF of the form

$$E(t) = \sum_{j=1}^{\mathcal{N}} E_j \cos(\omega t + \varphi_j). \tag{3.2.1}$$

For $\mathcal{N} \to \infty$ ($\mathcal{N}$ is the number of modes), the profile of a Stark component was obtained by *Lifshitz* [3.2]:

$$I_L(\Delta\omega) = \sum_{p=-\infty}^{+\infty} I_{|p|}(\tilde{\varepsilon}) \exp(-\tilde{\varepsilon}) \delta(\Delta\omega - p\omega), \quad \tilde{\varepsilon} \equiv (X_{\alpha\beta}\varepsilon)^2/2. \tag{3.2.2}$$

Here $I_{|p|}(\tilde{\varepsilon})$ are the modified Bessel functions; $\varepsilon = 3\hbar\bar{E}_0/(2m_e e\omega)$, $\bar{E}_0 = (\sum_{j=1}^{\mathcal{N}} E_j^2)^{1/2}$.

The most interesting is the strong modulation case ($\tilde{\varepsilon} \gg 1$) in which the form of the satellite envelope is essential. We use the asymptotic expansion of $I_{|p|}(\tilde{\varepsilon})$ [3.10]:

$$I_\nu(\nu z) = (2\pi\nu)^{-1/2}(1+z^2)^{-1/4}\exp(\nu\eta), \quad \nu \gg 1,$$
$$\eta \equiv (1+z^2)^{1/2} + \ln z/[1+(1+z^2)^{1/2}]. \tag{3.2.3}$$

Let $z \gg 1$. Then from (3.2.3) we find $I_\nu(\nu z) \approx (2\pi\nu z)^{-1/2}\exp[\nu z - \nu/(2z)]$. For the envelope of the satellites we obtain

$$\tilde{I}(p) \equiv I_{|p|}(\tilde{\varepsilon})\exp(-\tilde{\varepsilon}) \approx (2\pi\tilde{\varepsilon})^{-1/2}\exp[-p^2/(2\tilde{\varepsilon})], \quad \tilde{\varepsilon} \gg p \gg 1. \tag{3.2.4}$$

Now let $z \ll 1$. Then from (3.2.3) we have $I_\nu(\nu z) \approx (2\pi\nu)^{-1/2} \times \exp(\nu \ln z + \nu)$. For the envelope of the satellites we find

$$\tilde{I}(p) \approx (2\pi\tilde{\varepsilon})^{-1/2}\exp[-(p+1/2)\ln(p/\tilde{\varepsilon}) + p - \tilde{\varepsilon}], \quad p \gg \tilde{\varepsilon} \gg 1. \tag{3.2.5}$$

Thus $\tilde{I}(p)$ is described in a wide range $0 \leqslant p \ll \tilde{\varepsilon}$ by the Gauss law (3.2.4). The deviations from this law arise only at $p \gtrsim \tilde{\varepsilon}$ [i.e. at the intensity levels $\tilde{I}(p)/\tilde{I}(0) \lesssim \exp(-\tilde{\varepsilon}/2) \ll 1$] when the decrease of $\tilde{I}(p)$ is somewhat reduced.

We emphasize that $\tilde{I}(p)$, at fixed $\tilde{\varepsilon}$, has a maximum at $p = 0$ and $\tilde{I}(0) \approx (2\pi\tilde{\varepsilon})^{-1/2} = \pi^{-1/2}(X_{\alpha\beta}\varepsilon)^{-1}$. Hence, for multicomponent hydrogen SLs the intensity of the maximum (zeroth) satellite is equal to

$$\tilde{I}_0 = \left(f_0 + 2\sum_{k=1}^{k_{max}} f_k \right)^{-1} \left[f_0 + 2\pi^{-1/2}\varepsilon^{-1}\sum_{k=1}^{k_m}(f_k/X_k) \right], \tag{3.2.6}$$

here f_0 is the total intensity of the central components and f_k is the intensity of the kth lateral component. Two important corollaries follow from (3.2.6).

Corollary 1. *In the case $\langle X_k\rangle\varepsilon \gg 1$ for the SLs containing central components the halfwidth is in fact determined by the central components only. The lateral components may manifest only at the levels $I/I_{max} \lesssim (\langle X_k\rangle\varepsilon)^{-1}$.*

Corollary 2. *In the case $\langle X_k\rangle\varepsilon \gg 1$ for the SLs containing no central components, the component having the maximum value of f_k/X_k dominantly contributes to the intensity of the maximum. A profile of the satellite's envelope* may be calculated as a superposition of the Gauss profiles of the individual components.

The two-mode case. In the case $X_k\varepsilon \gg 1$ the satellite envelopes of *Blochinzew* [3.1] and *Lifshitz* [3.2] differ essentially: for the one component SL in a regular QEF $E_0 \cos \omega t$ the maximum is shifted by the distance $\Delta\omega_{max} \approx [X_k\varepsilon - 0.809(X_k\varepsilon)^{1/3}]\omega$, but upon the action of a turbulent QEF (3.2.1) the envelope is bell-shaped. Suppose that in some experiment a bell-shaped profile is really observed. Does this testify (as was asserted in [3.2]) to the existence of a QEF of the form of (3.2.1) with $\mathcal{N} \gg 1$? The answer is "no": the bell-shaped profile also arises for a small number of modes in (3.2.1), even at $\mathcal{N} = 2$.

In fact, the dependence of the distribution $W_{\mathcal{N}}(E_0)$ of the resulting amplitude E_0 on the mode number $\mathcal{N}$ is known from nonlinear optics [3.13]:

$$W_{\mathcal{N}}(E_0) = \frac{2(\mathcal{N} - 1)E_0}{\mathcal{N}\langle E_0^2\rangle}\left(1 - \frac{E_0^2}{\mathcal{N}\langle E_0^2\rangle}\right)^{\mathcal{N}-2}, \quad 0 \leqslant E_0 \leqslant (\mathcal{N}\langle E_0^2\rangle)^{1/2};$$

$$W_{\mathcal{N}}(E_0) = 0, \quad E_0 > (\mathcal{N}\langle E_0^2\rangle)^{1/2}. \tag{3.2.7}$$

At $\mathcal{N} \to \infty$, (3.2.7) transforms to the Rayleigh distribution $W(E_0) = (2E_0/\langle E_0^2\rangle) \exp(-E_0^2/\langle E_0^2\rangle)$. Consider the case of a two-mode field ($\mathcal{N} = 2$) which is the simplest intermediate case between the Blochinzew and Lifshitz models of QEFs. Averaging the Blochinzew correlation function $\Phi(\tau) = \sum_{p=-\infty}^{+\infty} J_p^2(X_{\alpha\beta}\varepsilon) \exp(ip\omega\tau)$ over the distribution $W_2(E_0)$ in (3.2.7), we obtain the corresponding profile of the Stark component

$$S_2\left(\frac{\Delta\omega}{\omega}\right) = \sum_{p=-\infty}^{+\infty} \delta\left(\frac{\Delta\omega}{\omega} - p\right) \tilde{I}_p(X_{\alpha\beta}\varepsilon),$$

$$\tilde{I}_p(y) \equiv J_p^2(y) - J_{p-1}(y)J_{p+1}(y). \tag{3.2.8}$$

We find the form of the satellite envelope in (3.2.8) for the case $X_{\alpha\beta}\varepsilon \equiv y \gg 1$. Using asymptotic representations of the Bessel functions [3.10] we obtain

$$\tilde{I}_p(y) \approx [2/(\pi y)][1 - (-1)^p(\cos 2y)/(2y)], \quad p \ll y;$$

$$\tilde{I}_p(y) \approx \left(\frac{2}{p}\right)^{4/3} f\left(-\left(\frac{2}{p}\right)^{1/3}(y - p)\right),$$

$$f(z) \equiv [Ai'(z)]^2 - zAi^2(z), \quad |p - y| \ll y. \tag{3.2.9}$$

The analysis of the dependence of $\tilde{I}_p(y)$ on p at fixed y leads to a bell-shaped envelope with a maximum at the line center ($p_{max} = 0$) and the halfhalfwidth $p_{1/2} \approx X_{\alpha\beta}\varepsilon$.

3.2.2 Numerical Calculations

The envelope of the Lifshitz satellites is described by the Gauss law (3.2.4) for the case $X_{\alpha\beta}\varepsilon \gg 1$ as was shown above. The Gauss distribution of the intensity of a hydrogen SL Stark component is also characteristic for the case where a one-dimensional multimode field

$$E(t) = \sum_{j=1}^{\mathcal{N}} E_j \cos(\omega_j t + \varphi_j) \tag{3.2.10}$$

acts on the atom quasistatically. This corresponds to the situation in which the lifetime of the excited atom τ_{life} is smaller than the oscillation periods of modes in (3.2.10) so that we may let $t = t_0$ =const: $E(t_0) \equiv F = \sum_{j=1}^{\mathcal{N}} E_j \cos \Phi_j$, $\Phi_j \equiv \varphi_j + \omega_j t_0$. If different terms here are statistically independent then according to the probabilistic laws the resulting field has a Gauss distribution

$$W_G(F) = (2/\pi)^{1/2} F_0^{-1} \exp[-F^2/(2F_0^2)], \quad F_0^2 = \int_0^\infty dF \, F^2 W_G(F).$$

$$\tag{3.2.11}$$

Thus, upon the action of one-dimensional BEFs (e.g. ion-acoustic turbulence) the Gauss distribution arises when $C_{\alpha\beta} F_0 \gg \tau_{\text{life}}^{-1} \gg \max_j \omega_j$. Upon the action of a multimode QEF (e.g. Langmuir turbulence) the Gauss distribution of intensities is relevant not only for the quasistatic limit $C_{\alpha\beta} F_0 \gg \tau_{\text{life}}^{-1} \gg \omega$ but also for the dynamic case of $C_{\alpha\beta} E_0 \gg \omega \gg \tau_{\text{life}}^{-1}$.

The use of the Gauss distribution instead of (3.2.2) allows us to create single-parameter tables of hydrogen-like SL profiles in the field (3.2.1). We have calculated the profiles $S_{t,nn'}^{(Z)}(\alpha)$, $S_{l,nn'}^{(Z)}(\alpha)$ corresponding to observation in directions transverse and along the field by the formula

$$S_{nn'}^{(Z)}(\alpha) = \left(\sum_{k=1}^{k_m} f_k\right)^{-1} \sum_{k=1}^{k_m} (Z^5 f_k/C_k)(2/\pi)^{1/2} \exp[-Z^{10}\alpha^2/(2C_k^2)],$$

$$\alpha \equiv \Delta\lambda/F_0, \tag{3.2.12}$$

where C_k values were tabulated in [3.14] (for the profiles $S_{l,nn'}^{(Z)}(\alpha)$ only σ-components should be taken into account). The calculations were carried out for hydrogen SLs ($Z = 1$) of the Balmer series ($n' = 2$, $n = 3$–8) as well as for ionized helium SLs ($Z = 2$) of the Fowler series ($n' = 3$; $n = 4$, $\lambda_0 = 468.6$ nm; $n = 5$, $\lambda_0 = 320.3$ nm; $n = 6$, $\lambda_0 = 273.3$ nm) and Pickering series ($n' = 4$; $n = 5$, $\lambda_0 = 1012.3$ nm; $n = 6$, $\lambda_0 = 656.0$ nm; $n = 7$, $\lambda_0 = 541.1$ nm). All calculated profiles have only one maximum, at $\alpha = 0$. The halfhalfwidths $\alpha_{1/2}^{(Z)}(n, n')$ of the calculated profiles defined from the relation $S(\alpha_{1/2}) = S_{\max}/2$ are presented in Appendix B.

Note that in spectroscopy of cosmic plasmas, along the line of observation there are many domains containing fields $E(t)$ of the type (3.2.1) but differing in the direction of the vector $E(t)$, these directions being uniformly distributed in the solid angle 4π. The corresponding calculated Balmer SL profiles H_n ($n = 3$–18) are contained in [3.15].

3.3 Impact Broadening of Hydrogen-like Spectral Lines

3.3.1 Modifications of Impact Broadening Due to QEF

1. Let the QEF $E(t) = E_0 \cos \omega t$ act upon a hydrogen atom in a plasma. Choose an axis Oz of a coordinate system $Oxyz$ along the vector E_0. Then the Schrödinger equation for a hydrogen atom interacting with the field $E(t)$ and the individual plasma microfields can be written (atomic units $\hbar = e = m_e = 1$ are used)

$$i \, \partial \Psi / \partial t = H\Psi, \quad H = H_a + zE_0 \cos \omega t + V(t),$$
$$V(t) = V_i + V_e(t), \tag{3.3.1}$$

where H_α is the unperturbed Hamiltonian of the hydrogen atom, $V_i = rF_i$ the operator of interaction with an ion microfield F_i, and $V_e(t) = rF_e(t)$ the operator of interaction with an electron microfield $F_e(t)$. We try the solution of (3.3.1) for a level with principal quantum number n in the basis of WFs in parabolic coordinates with an axis Oz, diagonalizing the operator $H_a + zE_0 \cos \omega t$:

$$\Psi(t) = \exp(-i\varepsilon_n t) \sum_{\alpha \in n} C_\alpha(t) \exp(-iz_{\alpha\alpha} E_0 \omega^{-1} \sin \omega t) \varphi_\alpha, \tag{3.3.2}$$

where $H_a \varphi_\alpha = \varepsilon_n \varphi_\alpha$, $z_{\alpha\alpha} \equiv \langle \varphi_\alpha | z | \varphi_\alpha \rangle$.

Substituting (3.3.2) into (3.3.1), we obtain

$$\dot{C}_\alpha = -i \sum_{\alpha' \in n} \sum_{r=-\infty}^{+\infty} \langle \varphi_\alpha | V(t) | \varphi_{\alpha'} \rangle C_{\alpha'} J_r(\Delta z_{\alpha\alpha'} E_0 \omega^{-1}) \exp(ir\omega t), \tag{3.3.3}$$

where $J_r(x)$ is a Bessel function, and $\Delta z_{\alpha\alpha'} \equiv z_{\alpha\alpha} - z_{\alpha'\alpha'}$. It is easy to see that the system (3.3.3) is equivalent to the Schrödinger equation with the Hamiltonian $H_0 + V(t)$, where the eigenvalues of the Hamiltonian H_0 are $E_{n,s} = \varepsilon_n + s\omega$ ($s = 0, \pm 1, \pm 2, \ldots$). Here the matrix elements of operator $V(t)$, calculated by WF $\Psi_{\alpha,s}$ of Hamiltonian $H_0 (H_0 \Psi_{\alpha,s} = E_{n,s} \Psi_{\alpha,s})$, are given by

$$\langle \Psi_{\alpha,s} | V(t) | \Psi_{\alpha',s-p} \rangle = \begin{cases} \langle \varphi_\alpha | V(t) | \varphi_{\alpha'} \rangle J_p(\Delta z_{\alpha\alpha'} E_0 \omega^{-1}), & \alpha \neq \alpha', \\ \langle \varphi_\alpha | V(t) | \varphi_\alpha \rangle \delta_{0,p}, & \alpha = \alpha', \end{cases} \tag{3.3.4}$$

where $\delta_{0,p}$ is the Kronecker symbol. Thus, the problem of the interaction of a hydrogen atom with the field $E(t)$ and individual plasma microfields can be reduced to the simpler problem of the interaction of an effective atom [hydrogen atom "dressed" by the field $E(t)$] with individual plasma microfields.

Consider the usual case in which an ion microfield leads to a quasistatic broadening of hydrogen lines, and the influence of an electron microfield to impact broadening. Assume that the frequency ω is larger than the splitting of the level n in an ion microfield F_i, and larger than the inverse time of an electron impact (Weisskopf frequency):

$$\omega \gg \max(n^2 F_i, v_e/\rho_{We}),\tag{3.3.5}$$

where v_e is an electron velocity, ρ_{We} is the Weisskopf radius. In this case the following procedure can be used for the calculation of the Stark broadening of hydrogen line satellites [we assume that the interaction with the field $E(t) = E_0 \cos\omega t$ has already been considered by introducing a dressed hydrogen atom]. At the first stage, eigenvalues and corresponding eigen-WFs of the static operator $H_0 + V_i$ are obtained (such a problem was considered in Chap. 2). Then the Stark profile of each satellite is determined by using an electron impact approximation at the fixed ion microfield [3.16]. For a calculation of such a profile one can use the electron impact broadening operator Φ_{ab}, which can be written as [3.16]

$$\Phi_{ab} = -(16e^4 N/3\hbar^2 \langle v \rangle)(r_a r_a + r_b^* r_b^* - 2r_a r_b^*)[\ln(\rho_m/\rho_{We}) + 0.215],\tag{3.3.6}$$

where $\rho_m = \min(\rho_{De}, \rho_{\Delta\omega})$, $\rho_{De} \equiv [kT/(4\pi Ne^2)]^{1/2}$, $\rho_{\Delta\omega} \equiv (\Delta\omega)^{-1}(2kT/m_e)^{1/2}$. While calculating matrix elements of Φ_{ab} it is necessary to take into account (3.3.4). Finally the resulting profiles of satellites are obtained by averaging over the probability distribution of ion microfields. It should be emphasized that in accordance with (3.3.4) the Stark width of each satellite of the SL is an oscillating function of the parameter E_0/ω, and in general the absolute value of this width is smaller than the width of the SL in the absence of the field $E_0 \cos\omega t$. These effects, which narrow the satellites, occur for both quasistatic and impact Stark broadening of hydrogen SLs.

Note that in typical conditions of experiments on the interaction of microwave radiation with low density plasmas ($N_e \lesssim 10^{13}$ cm^{-3}) the microwave frequency is $\omega \gg v_{Ti}/\rho_{Wi} \equiv \Omega_{Wi}(\Omega_{Wi}$ is the ionic Weisskopf frequency). In this case the broadening of QSs of hydrogen atoms is controlled mostly by ion (and not electron) impact broadening. So for the central Stark component of the L_α line the calculations show that the ion impact width γ_i is proportional to $N_e T_i^{-1/2} J_0^2(\mu)$, where $\mu = 3\hbar E_0/(m_e e\omega)$. Thus the halfwidth $2\Delta\omega_{1/2}$ of SL decreases with increasing microwave amplitude E_0. At some values of E_0, corresponding to the coincidence of μ with the zeros of the Bessel function J_0 ($\mu = 2.40, 5.52, \ldots$), the narrowing of the SL will be especially pronounced.

2. In the case of the QEF $E_0 \cos\omega t$ with parameters satisfying the inequality

$$\omega \ll \min[v_e/\rho_{We}, n^2 \hbar E_0/(m_e e)]\tag{3.3.7}$$

[which is in some sense opposite to the case of (3.3.5)] the electron impact broadening of hydrogen SL occurs, in fact, on a background of the static field $E = E_0 \cos \Phi$. It is well known that the standard semiclassical theories of impact broadening are intrinsically divergent at small impact parameteres [3.16–19]. Incorporation of a quasistatic field F into calculations of electron impact broadening in frames of these theories eliminated the divergence only at large impact parameters ρ [3.16–18]. The divergence at small ρ remained – finite results were obtained by a rather artificial cutoff at the Weisskopf radius. Due to this intrinsic divergency an inaccuracy of the results grows with an increase of the quasistatic field F.

We have developed a generalized semiclassical theory of Stark broadening that is free from that shortcoming – it is a convergent theory [3.20]. A Hamiltonian of a hydrogen atom under the action of a quasistatic field F and an electron-produced dynamic field $E(t)$ may be written in the form

$$H \equiv H_0 - dF - dE(t) \tag{3.3.8}$$

where H_0 is an unperturbed Hamiltonian, d – a dipole moment operator. Choose the axis Oz of parabolic quantization along the field F.

A physical idea behind our approach is that the interaction $-d_z(F + E(z))$ with an entire z-component of the total field (and not only its part $-d_z F$) is diagonal in any n-subspace. Therefore the z-component of the electron microfield may be allowed for more accurately than in the standard theories where the entire interaction $-dE(t)$ was treated by the time-dependent perturbation theory.

In spite of much more complicated starting formulas, we have managed to develop this theory *analytically* to the same level as the standard theories. Our net result may be presented as a substitution of the "broadening" function $C_{ST}(Z)$ of the standard theories (that enters the electron impact broadening operator $\Phi_{\alpha\beta}$) by a generalized but still elementary function $C(\chi, Y, Z)$:

$$C(\chi, Y, Z) \equiv (3/4) \int_{-\infty}^{+\infty} dx_1 \int_{-\infty}^{x_1} dx_2 \exp[iZ(x_1 - x_2)][(1 + x_1^2)(1 + x_2^2)]^{-3/2}$$

$$\times \{\varepsilon^{-1} \sin \varepsilon + (2x_1 x_2 - 1)\varepsilon^{-1} j_1(\varepsilon)$$

$$+ (Y/Z)^2[x_1(1 + x_1^2)^{-1/2} - x_2(1 + x_2^2)^{-1/2} + 2\chi]$$

$$\times [x_1(1 + x_2^2)^{1/2} - x_2(1 + x_1^2)^{1/2} - 2\chi(x_1 x_2 - 1)]\varepsilon^{-2} j_2(\varepsilon)\},$$

$$\varepsilon \equiv 2^{1/2}(Y/Z)\{1 - (1 + x_1 x_2)[(1 + x_1^2)(1 + x_2^2)]^{-1/2}$$

$$- \chi[x_2(1 + x_2^2)^{-1/2} - x_1(1 + x_1^2)^{-1/2}] + 2\chi^2\}^{1/2}. \tag{3.3.9}$$

Variables x_1, x_2, Z in (3.3.9) are notations of the standard theories

$$x_1 \equiv vt_1/\rho, x_2 \equiv vt_2/\rho,$$

$$Z_k \equiv (\delta d)_k \rho F/(\hbar v); k = \alpha, \beta$$

$$(\delta d)_k \equiv (d_z)_{kk} - (d_z)_{k'k'} \tag{3.3.10}$$

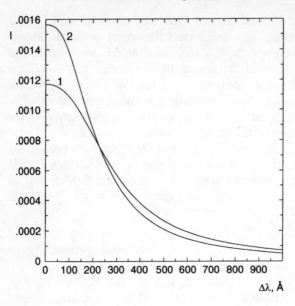

Fig. 3.9. Profiles of the H_β line calculated at $N_e = N_i = 10^{18}$ cm^{-3}, $T_e = 1$ eV and a quasistatic field strength $F = 600$ kV/cm: 1 – standard theory, 2 – generalized theory from [3.20]

There are two new parameters in the function $C(\chi, Y, Z)$. The first one χ stands for

$$\chi_\alpha = [\text{sign}(\delta d)_\alpha] X_{\alpha\beta}/n \text{ or } \chi_\beta = [\text{sign}(\delta d)_\beta] X_{\alpha\beta}/n', \quad X_{\alpha\beta} = nq - n'q';$$

$$q = n_1 - n_2, \quad q' = n_1 - n_2'; \quad \alpha \equiv (n_1 n_2 m), \quad \beta \equiv (n_1' n_2' m'). \tag{3.3.11}$$

The second new parameter Y is

$$Y_k \equiv (\delta d_z)_k^2 eF/(\hbar v)^2$$

$$= [3n_k\hbar/(2m_e v)]^2 F/e; k = \alpha, \beta. \tag{3.3.12}$$

Our theory embraces the standard theories as a limiting case $Y \to 0$ corresponding to $F \to 0$. The higher the strength of a low-frequency QEF and/or the ion density, the greater this parameter Y becomes as well as the inaccuracy of the standard theories.

Figure 3.9 shows profiles of the H_β-line calculated at $N_e = N_i = 10^{18}$ cm^{-3}, $T_e = 1$ eV for the quasistatic field strength $F = 600$ kV/cm. Note that $F = 600$ kV/cm numerically corresponds to the most probable value of ion microfield $F = 4.2 \, e N_i^{2/3}$ at $N_i = 10^{18}$ cm^{-3}. It is seen from the Fig. 3.9 that the standard theory overestimates broadening by 33%.

3.3.2 Modification of QEF-induced Line Splitting Due to Impact Broadening

Consider the profile $I_{q.s.}(\Delta\omega)$ of a Stark component of a hydrogen SL split by a QEF $E_0 \cos \omega t$ in the quasistatic limit. In this case, the individual emitter experiences the action of a stationary field $E_0 \cos \Phi$ ($\Phi = \omega t_0$) so that

$$I_{\text{q.s.}}(\Delta \omega) = (2\pi)^{-1} \int_0^{2\pi} d\Phi \, \delta(\Delta \omega - CE_0 \cos \Phi)$$

$$= \pi^{-1}[(CE_0)^2 - (\Delta \omega)^2]^{-1/2}, \tag{3.3.13}$$

where $C \equiv 3X_{\alpha\beta}\hbar/(2Zm_e e)$. The profile (3.3.13) has singularities at the points $\Delta \omega = \pm CE_0$ [see also (3.1.6)].

In general, such singularities appear in calculation of the following integrals:

$$I = \int dx \, g(x) \, \delta(f(x)),$$

$$f(x_0) = f'(x_0) = 0, \qquad g(x_0) \neq 0, \qquad a < x_0 < b \tag{3.3.14}$$

so that in the standard formula

$$I = \sum_{k=0, 1, \dots} g(x_k)[|f'(x_k)|]^{-1}, \qquad f(x_k) = 0 \tag{3.3.15}$$

the term with $k = 0$ has no sense. In this situation, from the physical point of view, to obtain a finite value of I it is necessary to replace the δ-function by a real profile, e.g. by the dispersive profile

$$\delta(f(x)) \approx \pi^{-1} \frac{\gamma}{\gamma^2 + f^2(x)}\bigg|_{x \approx x_0} \approx \pi^{-1} \frac{\gamma}{\gamma^2 + [f''(x_0)/2]^2(x - x_0)^4}. \tag{3.3.16}$$

Then, using the integral $\pi^{-1} \int_{-\infty}^{+\infty} dz \, \alpha/(\alpha^2 + z^4) = (2\alpha)^{-1/2}$ we find

$$I \approx g(x_0)[\gamma|f''(x_0)|]^{-1/2} + \sum_{k=1, 2, \dots} g(x_k)[f'(x_k)]^{-1}. \tag{3.3.17}$$

Returning to the plasma spectroscopy problem, we emphasize first of all that replacing the profile $\delta(\Delta \omega - CE_0 \cos \Phi)$ by the dispersive one corresponds to allowing for the impact broadening by charged plasma particles. Using (3.3.17) we obtain

$$I_{\text{q.s.}}(\pm CE_0) \approx \pi^{-1}(\gamma CE_0/2)^{-1/2}. \tag{3.3.18}$$

The condition of validity for the quasistatic description of the profile may now be represented in the form

$$[\omega/(CE_0)]^{2/3} \ll \gamma/(CE_0) < 1. \tag{3.3.19}$$

In the opposite (dynamic) case, in which $\gamma \ll (CE_0)^{1/3}\omega^{2/3}$, the profile $I_B(\Delta \omega)$ at the points $\Delta \omega = \pm CE_0$ is equal to [see (3.1.12)]

$$I_B(\pm CE_0) \approx 0.911(CE_0)^{-2/3}\omega^{-1/3}. \tag{3.3.20}$$

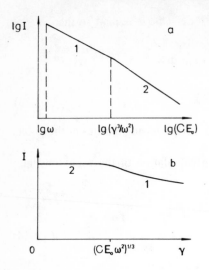

Fig. 3.10. Maximum intensity of a Stark component profile of a hydrogen-like SL vs. QEF amplitude E_0 at fixed QEF frequency ω and impact width γ (**a**) and vs. γ at fixed E_0 and ω (**b**): *Curve 1* – quasistatic region, *2* – dynamic region. In **a** the slopes q of the straight lines are $q = -1/2$ for region 1 and $q = -2/3$ for region 2. In **b** for region 1, $I \propto \gamma^{-1/2}$

Equations (3.3.18, 20) match each other well at the border of their regions of validity.

The function of the maximum intensity of the Stark component profile vs. E_0 (at fixed ω, γ) and vs. γ (at fixed E_0, ω) for the quasistatic and dynamic cases are shown in Fig. 3.10.

Thus, in the situation in which the quasistaticity of QEF action is caused by impact broadening (by charged particles) the maxima and the halfwidth of the whole SL $I_{q.s.}(\Delta\omega)$ are determined by the maxima and the halfwidth of the following *effective* profile, [see (3.1.12)],

$$I^{\text{eff}}_{q.s.}(\Delta\omega) = \frac{1}{f_0 + 2\sum\limits_{k=1}^{k_m} f_k} \left[f_0 \frac{\pi^{-1}\gamma_0}{\gamma_0 + (\Delta\omega)^2} + \sum_{k=1}^{k_m} \frac{f_k}{\pi(\gamma_k C_k E_0/2)^{1/2}} \right.$$

$$\times \left. \left(\frac{1}{1 + (\Delta\omega - C_k E_0)^2/\gamma_k^2} + \frac{1}{1 + (\Delta\omega + C_k E_0)^2/\gamma_k^2} \right) \right],$$
$$(3.3.21)$$

where γ_k is the impact width of kth Stark component (calculated, e.g. in [3.18]). Two important corollaries follow from (3.3.21).

Corollary 1. *For lines containing central components ($f_0 \neq 0$) the ratio of intensities of the central and lateral maxima is equal to $(\gamma_k C_k E_0/2)^{1/2} f_0/(\gamma_0 f_k)$ $\gg 1$ so that the profile halfwidth is $\Delta\omega_{1/2} \approx 2\gamma_0$.*

Corollary 2. *For lines containing no central components ($f_0 = 0$) the effective profile reproduces the picture of the static Stark splitting in the field E_0 but with redefined component intensities, see (3.1.13)*

$$f^{\text{eff}}_k \equiv f_k \cdot (X_k \gamma_k)^{-1/2}.$$
$$(3.3.22)$$

This may lead to changes in the positions of the maxima and the line halfwidth in comparison to the static Stark profile.

3.4 Frequency-integrated Radiative Characteristics of Hydrogen-like Emitters Interacting with a Resonant Laser Field and a Low-Frequency QEF

3.4.1 Resonant Multiquantum Interaction

Let us investigate a resonant multiquantum interaction of an atom with a multifrequency EF of the type

$$E(t) = E_{01} \cos(\omega_1 t + \theta) + E_{02} \cos \omega_2 t, \quad \omega_2 \ll \omega_1, \tag{3.4.1}$$

when a field $E_{01}(t) \cos(\omega_1 t + \theta)$ is resonant with an atomic transition frequency $\omega_{ab}^{(0)}$ ($|\omega_1 - \omega_{ab}^{(0)}| \ll \omega_{ab}^{(0)}$). The problem is to find frequency-integrated quantities such as an induced transition probability (per unit time) between lower (b) and upper (a) levels, the stationary difference of level populations, the power absorbed by a medium, etc. These quantities are directly determined by the squared modulus of the compound matrix element v_{ba} [3.21, 22].

Consider the situation in which a field $E_{02} \cos \omega_2 t$ strongly affects atomic states so that perturbation theory is inapplicable. In this case the calculation of v_{ba} may be subdivided into two steps. Firstly, the WFs $\Psi_a(t)$, $\Psi_b(t)$ of QSs of the levels a, b in a field $E_{02} \cos \omega t$ are determined. Secondly, a matrix element of a transition between QS $\Psi_a(t)$, $\Psi_b(t)$ is calculated assuming a *one-photon* resonance with a field $E_{01} \cos(\omega_1 t + \theta)$.

3.4.2 Non-Degenerate Case

Let the dipole moments $d_{\alpha\alpha}$ and $d_{\beta\beta}$ exist in *nondegenerate* atomic states (e.g. Stark sublevels of hydrogen in a static EF). We represent the multiquantum resonance condition in the form

$$\omega_1 + r\omega_2 = \omega_{\alpha\beta}^{(0)} + \delta; \quad r = 0, \pm 1, \pm 2, \ldots; \quad |r|\omega_2 \ll \omega_1, \quad |\delta| \ll \omega_2. \tag{3.4.2}$$

Then a matrix element $v_{\beta\alpha}$ is expressed by the formula ($\hbar = m_e = e = 1$)

$$v_{\beta\alpha} = 2^{-1}(v_1)_{\beta\alpha} J_r(\Delta v_2/\omega_2) \exp(i\theta);$$
$$v_1 = -dE_{01}, \quad v_2 = -dE_{02}, \quad \Delta v_2 = (v_2)_{\beta\beta} - (v_2)_{\alpha\alpha}, \tag{3.4.3}$$

where $J_r(x)$ is a Bessel function. Substituting (3.4.3) into the expression for a stationary difference of populations $\eta_s = (\bar{\sigma}_{\beta\beta} - \bar{\sigma}_{\alpha\alpha})_s$ of resonating levels [3.21, 22] we find

$$\eta_s = \eta_0 \left\{ 1 + G_r \left[1 + (\omega_{\alpha\beta}^{(0)} - \omega_1 - r\omega_2)^2 T^2 \right]^{-1} \right\}^{-1}. \tag{3.4.4}$$

Here $\eta_0 = (\bar{\sigma}_{\beta\beta} - \bar{\sigma}_{\alpha\alpha})_0$ is the initial difference of level populations; the saturation parameter G_r is equal to

$$G_r = \tau T \left|(v_1)_{\beta\alpha}\right|^2 J_r^2(\Delta v_2/\omega_2), \tag{3.4.5}$$

where τ and T are times of longitudinal and transverse relaxation. Using (3.4.4) we determine the power Q_j which is absorbed (at $Q_j > 0$) or emitted (at $Q_j < 0$) by a unit volume at a frequency $\omega_j (j = 1, 2)$

$$Q_j = (2\tau)^{-1} N n_j \omega_j (\eta_0 - \eta_s), \tag{3.4.6}$$

where $n_1 = 1$, $n_2 = r$, N is the number of atoms per unit volume. The total power per unit volume of medium is equal to

$$Q = Q_1 + Q_2 = (2\tau)^{-1} N \omega_{\alpha\beta}^{(0)} (\eta_0 - \eta_s)$$

$$= (2\tau)^{-1} N \omega_{\alpha\beta}^{(0)} G_2 \eta_0 \left[1 + (\omega_{\alpha\beta}^{(0)} - \omega_1 - r\omega_2)^2 T^2 + G_r\right]^{-1}. \tag{3.4.7}$$

(For the particular case of $r = 0$ an analogous formula was obtained in [3.23].) It follows from (3.4.6,7) that the spectrum of absorption of a field $E_{01} \cos(\omega_1 t + \theta)$ consists of a set of satellites, the halfwidth of a satellite of index r being equal to $\Delta\omega_{1/2}^{(r)} = 2T^{-1}(1 + G_r)^{1/2}$. In the case $G_r \gg 1$, the halfwidth is $\Delta\omega_{1/2}^{(r)} = 2(\tau/T)^{1/2}|(v_1)_{\alpha\beta} J_r(\Delta v_2/\omega_2)|$. Usually $T \ll \tau$, therefore when $4^{-1}(T/\tau)^{1/2} < |v_{\beta\alpha}|/\omega_2 \ll 1$, the satellite halfwidth is $\Delta\omega_{1/2}^{(r)} \gg \omega_2$. In this case spectra of individual satellites overlap and an absorption band arises, with a halfwidth $\Delta\omega_{1/2}^{band} \gg \omega_2$. At fixed $E_{01}, \omega_1, \omega_2$ the functions $G_r(E_{02})$ (3.4.5) and $Q(E_{02})$ (3.4.7) are oscillatory. Hence, it is possible to come in and out of saturation many times by a monotonic variation of E_{02}.

Note that for the values of E_{02} at which $J_r(\Delta v_2/\omega_2) = 0$, a supertransparency of a medium occurs. This transparency enhancement is caused not by the usual saturation effect but by the vanishing of the matrix element $v_{\beta\alpha}$.

Let us also analyze (3.4.6) for the power $Q_2^{(r)}$ absorbed (or emitted) by a medium at a frequency ω_2. At saturation we have $Q_2^{(r)} \propto r J_r^2(\Delta v_2/\omega_2)$. Therefore in weak fields $|\Delta v_2/\omega_2| < 1$ the function $|Q_2^{(r)}|$ reaches a maximum at $|r| = 1$. However, at $|\Delta v_2/\omega_2| \gg 1$ the maximum of $Q_2^{(r)}$ is reached at $|r| > 1$; a considerable increase in $\max_r |Q_2^{(r)}|$ occurs with an increase in $|\Delta v_2/\omega_2|$. Thus, by tuning to resonances with different r in (3.4.2) it is possible to regulate absorption (or enhancement) of a low-frequency field $E_{02} \cos \omega_2 t$ without changing the laser field amplitude E_{01}.

3.4.3 Degenerate Case

The following results are obtained for a transition between *degenerate* states a and b of a Coulomb radiator. In particular, at $E_{01} \| E_{02}$ an absorbed (or emitted) power Q per unit volume may be written

$$Q = \frac{\omega_{ab} G_r (N_a^{(0)}/g_a - N_b^{(0)}/g_b)}{2\tau'[1 + (\omega_{ab} - r\omega_2 - \omega_1)^2 T^2 + G_r]}; \qquad \tau' \equiv \frac{(g_a + g_b)\tau}{2 g_a g_b},$$

$$G_r = 4\tau' T \Omega_\Sigma^2(r), \qquad \Omega_\Sigma^2(r) \equiv \frac{E_{01}^2}{4} \sum_{\alpha \in a, \ \beta \in b} z_{\alpha\beta}^2 J_r \left(\frac{(z_{\alpha\alpha} - z_{\beta\beta}) E_{02}}{\omega_2} \right).$$

$$(3.4.8)$$

Here N_a, N_b are initial concentrations of atoms in the states a, b; g_a, g_b are the statistical weights of states a, b; τ is the lifetime of the upper level. Formula (3.4.8) is valid under the usual assumption about the presence of rapid internal cross-relaxation processes sustaining equal populations of all states of the same level (a or b) during a laser pulse. Thus, for example, for a hydrogen atom, the estimations show that $T \ll \tau' \sim (10^{-9} - 10^{-10})$ s at electron densities $N_e \gtrsim 10^9$ cm^{-3}.

We emphasize that (3.4.8) is valid at time intervals $\Delta t \gg \max(\tau', T)$ [and similarly (3.4.7) at $\Delta t \gg \max(\tau, T)$]. In the opposite case, $\Delta t \ll \min(\tau', T)$, the evolution of a Coulomb radiator interacting with a field of type (3.4.1) may be described in terms of generalized QSs. In [3.24] for the H_α line the quasienergies of such QSs (and their Rabi frequencies) were found. It is interesting to note that the value $\Omega_\Sigma^2(r)$ [to which the saturation parameter in (3.4.8) is proportional] turns out to be equal to the sum of squares of all arising Rabi frequencies for the transition $a \leftrightarrow b$ at zero detuning (Appendix B).

The important feature of the considered problems is that for all hydrogen-like SLs (except the Lyman series) many quasienergy harmonics of upper and lower levels are simultaneously involved into a resonance (*multifrequency resonance*). Indeed at $n_b \neq 1$ the resonance condition is fulfilled for any harmonic q of lower level b:

$$\omega_1 \approx E_a^{(0)} + r\omega_2 - E_b^{(0)}$$
$$= [E_a^{(0)} + (r + q)\omega_1] - (E_b^{(0)} + q\omega_1), \qquad q = 0, \pm 1, \dots . \qquad (3.4.9)$$

The multifrequency character of the resonance was taken into account when obtaining (3.4.5,8) by using the sum rules

$$\sum_{k=-\infty}^{+\infty} J_k(z_{\alpha\alpha} E_{02} \omega_2^{-1}) J_{k-r}(z_{\beta\beta} E_{02} \omega_2^{-1}) = J_r((z_{\alpha\alpha} - z_{\beta\beta}) E_{02} \omega_2^{-1}). \qquad (3.4.10)$$

3.4.4 Applications of the Results

The results obtained can be the basis of a method for measurements of parameters of strong QEFs (e.g. in the microwave range) in plasmas. The essence of the method is as follows. Under the action of laser radiation at a frequency ω_1 satisfying the resonance conditions (3.4.2) with $\delta = 0$, enhancement of a population $\Delta \bar{\sigma}_{aa}$ of the upper level a occurs:

$$\Delta \bar{\sigma}_{aa} = 2^{-1}(\eta_0 - \eta_s) = \eta_0 G_k(E_{02})\{2[1 + G_k(E_{02})]\}^{-1}. \qquad (3.4.11)$$

For measurements of E_{02} one should detect the frequency-integrated intensity of a spontaneous transition (fluorescence signal) $I_{fl} \propto \Delta \bar{\sigma}_a$ from the level a to one of the lower levels, versus laser power $I_L \propto E_{01}^2$:

$$I_{fl}^{-1} \propto (1 + I_{sat}^{(k)} I_L^{-1}), \qquad I_{sat}^{(k)} I_L^{-1} \equiv G^{-1}(E_{02}). \tag{3.4.12}$$

Measurements of E_{02} in plasmas can be carried out in two ways. In the first, one tunes the laser frequency ω_1 to the resonance (3.4.2), measures the ratio of the slopes of $I_{fl}^{-1}(I_L^{-1})$ for some known value of E_{02}^* (e.g. for $E_{02}^* = 0$ at $r = 0$) and for a sought value E_{02} and uses the theoretical dependence of G on E_{02}. In Fig. 3.11, as an example, theoretical dependences of the reduced saturation parameter $g_k \equiv I_{sat}^{(0)}(0)[I_{sat}^{(k)}(E_{02})]^{-1}$ for the SL H_α and H_β are shown. Realization of this method for local measurements of a QEF amplitude (namely, a microwave amplitude) is discussed in Chap. 7.

In the second method one measures the ratio of the slopes of $I_{fl}^{-1}(I_L^{-1})$ obtained at the same E_{02} but at different $r(r = r'$ and $r = r'')$ and uses the dependence of $G_{r'}/G_{r''}$ on E_{02}. In other words, one should tune to frequencies of the unperturbed line and to a satellite or to frequencies of two different satellites. In particular, for SLs H_α and H_β the theoretical dependence of $G_{\pm1}/G_0$ on E_{02} is easily determined from the ratio of ordinates of the dashed and solid curves in Fig. 3.11.

We emphasize that (3.4.7,8) represent a nontrivial generalization of the well-known [3.21,22] Karplus–Schwinger formula for absorbed laser power to the case when an additional low-frequency field $E_{02} \cos \omega_2 t$ is present. Naturally these formulas may be used for measurements of an E_{02} not only by a fluorescence signal but also directly by laser absorption.

The results presented in this section can also be implemented in laser physics. Consider as an example a laser whose active medium consists of

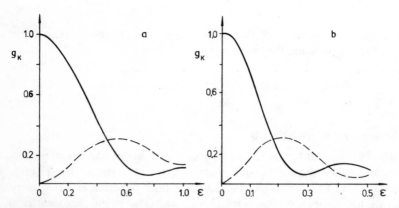

Fig. 3.11. The dependence of the reduced saturation parameter $g_k(\varepsilon) = I_{sat}^{(0)}(0) \times [I_{sat}^{(k)}(E_{02})]^{-1}$ vs. a reduced amplitude $\varepsilon = 3\hbar E_{02}/2Zm_e e\omega_2$ of a low-frequency field at $k = 0$ (*solid curve*) and at $k = \pm1$ (*dashed curve*): (**a**) for the H_α line; (**b**) for the H_β line

atoms possessing dipole moments in the states α, β. Let a microwave field $E_{02} \cos \omega_2 t$ act on such a medium. In accordance with the resonance condition of (3.4.2) the laser has a set of amplification coefficients $\gamma_{\omega_1}^{(r)}$ at frequencies $\omega_1 \approx \omega_{\alpha\beta}^{(0)} - r\omega_2 (r = 0, \pm 1, \pm 2, \ldots)$. In accordance with (3.4.5) the coefficients are proportional to squares of effective matrix elements of the dipole moments:

$$\gamma_{\omega_1}^{(r)} \propto \left| d_r^{\text{eff}} \right|^2 \equiv |d_{ba}|^2 \, J_r(\Delta v_2/\omega_2); \quad r = 0, \pm 1, \ldots. \tag{3.4.13}$$

Hence the frequencies of the laser may be retuned by changing the value of E_{02}. Indeed, with an increase of E_{02} a threshold inversion may be reached at a value of $r \neq 0$ such that $J_r^2(\Delta v_2/\omega_2) > J_{r'}^2(\Delta v_2/\omega_2)$ for all $r' \neq r$. The amplification coefficient $\gamma_{\omega_1}^{(r)}$ and consequently the output laser power at the frequency $\omega_1 \approx \omega_{\alpha\beta}^{(0)} - r\omega_2$ may also be retuned by changing E_{02}.

By using a microwave field on an active medium it is possible to generate a giant pulse without modulation of the Q-factor of the optical resonator. Upon switching on a strong microwave field, a large difference in the populations of levels α, β is created at the expense of a considerable decrease of all amplification coefficients $\gamma_{\omega_1}^{(r)}$ (in order that lasing cannot arise at any r). When the microwave field is switched off fast, a high intensity pulse is generated.

4 Action of Multidimensional Dynamic Electric Fields on Coulomb Emitters

This chapter is mostly devoted to dynamic resonances in atomic (or ionic) spectra. In contrast to the usual resonance case in which a dynamic field resonates with an unperturbed (or slightly perturbed) atom, a dynamic resonance occurs only when the separation between the energy levels is strongly changed by the dynamic field. The second distinctive feature of the dynamic resonance is its multifrequency nature (in spite of the fact that the dynamic field is single-mode). In the last section of this chapter the opposite case is treated in which a dynamic field is too high-frequency or too strong to produce resonances [4.1–7].

4.1 Splitting of Hydrogen-like Spectral Lines in a Plane Polarized QEF

4.1.1 Analytical Results for a Circularly Polarized Field

The problem considered in this section was first solved in [4.8,9]. We discuss the results only briefly since the solution was also published in a review [4.10] and also because it represents a particular case of a more general problem to which Sects. 4.1.2–4 are devoted.

If an EF of strength E_0 rotates with a constant angular frequency $\dot{\varphi} = \omega$, then in the reference frame rotating with the same frequency the Schrödinger equation is written (here and below $\hbar = m_e = e = 1$):

$$i\,\partial\Phi/\partial t = H\Phi, \quad H = H_a + zE_0 - \omega l_y, \quad \Phi(t) = \exp(i\omega t l_y)\Psi(t).$$
(4.1.1)

Here l_y is the projection operator of orbital angular momentum, $\Psi(t)$ is a WF in the rest reference frame. The relative strengths of "magnetic" $(-\omega l_y)$ and electric (zE_0) interactions may be conveniently characterized by the parameter $X = 2Z\omega/3nE_0$, where n is the principal quantum number and Z is the nuclear charge of the radiating particle.

The exact solution of the problem uses the additional integral of motion in a Coulomb field which is known as the Runge–Lenz vector A [4.11]. This allows to represent the Hamiltonian in (4.1.1) as

$$H = H_a + J_1\omega_1 + J_2\omega_2; \quad J_{1,2} \equiv (l \pm A)/2, \quad \omega_{1,2} \equiv \omega \pm [3n/(2Z)]E_0.$$
(4.1.2)

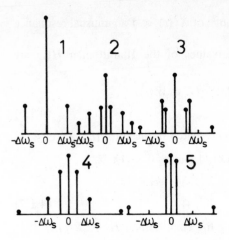

Fig. 4.1. The spectrum of the L_α line in a circularly polarized field: $1 - X = 0$; $2 - X = 0.2$; $3 - X = 0.5$; $4 - X = 1.0$; $5 - X = 2.0$

The WFs $u_{nn'n''}$, which diagonalize the Hamiltonian in (4.1.2), correspond to the definite projection of J_1 on ω_1 (denoted by a quantum number n') and to the definite projection of J_2 on ω_2 (denoted by n''). They are obtained from the usual parabolic WFs $u_{ni_1 i_2}$ (i_1, i_2 are quantum numbers of the projections of J_1 and J_2 on E_0) by rotations to angles β_1 and β_2 formed by the vectors ω_1 and ω_2 with the vector E_0. Eigenvalues of perturbed and unperturbed Hamiltonians differ by the value

$$W = (n' + n'')|\omega_{1,2}| = (n' + n'')(1 + X^2)^{1/2}\Delta\omega_s, \quad \Delta\omega_s \equiv 3nE_0/(2Z).$$
(4.1.3)

As an example, in Fig. 4.1 the spectra of SL L_α (taken from [4.8]) are shown.

4.1.2 Multiquantum Dynamic Resonance in an Elliptically Polarized Field

The Schrödinger equation may be represented in the form

$$i\,\partial\Psi/\partial t = \tilde{H}\Psi, \quad \tilde{H} = H_a + x\xi\varepsilon_0\sin\omega t + z\varepsilon_0\cos\omega t,$$
$$\varepsilon_0 \equiv E_0(1 + \xi^2)^{-1/2},$$
(4.1.4)

where E_0 is the field intensity and ξ is the ellipticity degree (ED). The electric field $E(t)$ rotates with the frequency $\dot\varphi(t) = \xi\omega/(\cos^2\omega t + \xi^2\sin^2\omega t)$. In the coordinate system rotating with the same frequency, from (4.1.4) we obtain

$$i\,\partial\Phi/\partial t = H\Phi, \quad H_a = zE(t) - \dot\varphi(t)l_y;$$
$$\Phi(t) \equiv \exp[i\varphi(t)l_y]\Psi(t), \quad E(t) \equiv \varepsilon_0(1 - k^2\sin^2\omega t)^{1/2}, \quad k^2 \equiv 1 - \xi^2.$$
(4.1.5)

It is convenient to describe the relative significance of the "magnetic" $(-\dot\varphi l_y)$ and electric interactions by the parameter $X(t) \equiv 2Z\dot\varphi(t)/[3nE(t)]$. It is shown below for the case of $n = 2$ that in the range of $X^2(t) \ll 1$ an unusual resonance effect may occur.

For $n = 2$ the instantaneous eigenvalues of the Hamiltonian $H(t)$ are [4.8,9]:

$$W_1(t) = -W_4(t) = 3E(t)[1 + X^2(t)]^{1/2}/Z \equiv W(t),$$
$$W_2(t) \equiv W_3(t) \equiv 0. \tag{4.1.6}$$

The corresponding WFs $\psi_j(t)$ may be represented in the form [for $X^2(t) \ll 1$]

$$\tilde{\psi}_1 - \psi_1 = \tilde{\psi}_4 - \psi_4 = -(\psi_1 + \psi_4)X^2/4 + i(\psi_2 + \psi_3)X/2,$$
$$\tilde{\psi}_2 = i(\psi_1 + \psi_4)X/2 + \psi_2(1 - X^2/4) - \psi_3 X^2/2,$$
$$\tilde{\psi}_3 = i(\psi_1 + \psi_4)X/2 + \psi_3(1 - X^2/4). \tag{4.1.7}$$

Here ψ_j are parabolic WFs diagonalizing the Hamiltonian $H_a + zE(t)$:

$$\psi_1 = |100\rangle, \quad \psi_2 = |001\rangle, \quad \psi_3 = |00-1\rangle, \quad \psi_4 = |010\rangle.$$

We try to solve (4.1.5) using the adiabatic basis

$$\Phi(t) = \sum_{j=1}^{4} c_j(t)\tilde{\psi}_j(t) \exp\left[-i\int_0^t dt'\, W_j(t')\right]. \tag{4.1.8}$$

Substituting (4.1.8) into (4.1.5) we obtain the set of equations

$$\dot{c}_{1,4} = -(i/2)\dot{X}(c_2 + c_3)\,\exp(\pm i\beta),$$
$$\dot{c}_2 = \dot{c}_3 = -(i/2)\dot{X}[c_1\exp(-i\beta) + c_4\exp(i\beta)],$$

$$\dot{X} \equiv \xi Zk^2\omega^2(\sin 2\omega t)/[2\varepsilon_0(1 - k^2\sin^2\omega t)^{5/2}], \quad \beta(t) \equiv \int_0^t dt'\, W(t'),$$
$$\tag{4.1.9}$$

(here and below the first subscripts of $\dot{c}_{1,4}$, $c_{1,4}$ correspond to the upper sets of signs, and the second subscripts to the lower ones). We expand $X(t)$ and $W(t)$ in the Fourier series:

$$\dot{X}(t) = \sum_{q=-\infty}^{+\infty} u_q\exp(2iq\omega t), \quad u_q \equiv -i\xi Zk^2\omega^2(g_{|2q-2|} - g_{|2q+2|})/(8\varepsilon_0),$$

$$W(t) \approx (b_0/2 + \sum_{p=1}^{\infty} b_{2p}\cos 2p\omega t)3\varepsilon_0/Z + (g_0/2$$

$$+ \sum_{q=1}^{\infty} g_{2q}\cos 2q\omega t)Z\xi^2\omega^2/(6\varepsilon_0), \tag{4.1.10}$$

where

$$g_{2q} = 2(-1)^q \sum_{r=q}^{+\infty}(k/2)^{2r}(2r+3)!!(2r-1)!!/[3(r-q)!(r+q)!],$$

$$b_{2p} = 2(-1)^{p+1} \sum_{r=p}^{+\infty} (k/2)^{2r} (2r-3)!!(2r-1)!!/[(r-p)!(r+p)!],$$

$$(p \geqslant 2),$$

$$b_0 = 4/[\pi \mathbb{E}(k)], \qquad b_2 = [4/(3\pi)][\mathbb{E}(k) - 2(1-k^2)\mathbb{D}(k)], \qquad (4.1.11)$$

where $\mathbb{E}(k)$, $\mathbb{D}(k)$ are complete elliptic integrals.

If we formally suppose $X(t) \equiv 0$, then for the solutions of (4.1.5) the QSs with the following WFs may be chosen:

$$\Phi_j = \tilde{\psi}_j \exp\left[-i \int_0^t dt' \, W_j(t')\right], \qquad (j = 1, 2, 3, 4). \qquad (4.1.12)$$

The quasienergy separations Q of these QSs are expressed through the time averaged instantaneous splitting $W_n(t)$ in crossed electric and "magnetic" fields

$$Q = \langle W_n(t) \rangle + 2r\omega \quad (r = 0, \pm 1, \ldots),$$

$$\langle W_n(t) \rangle \approx 3n\varepsilon_0 b_0/Z + Z\xi^2\omega^2 g_0/6n\varepsilon_0. \qquad (4.1.13)$$

Here and below, the formulas explicitly containing the principal quantum number n express the more general results (valid for any n).

In reality the value $\dot{X}(t) \not\equiv 0$ acts as the perturbation which can give rise to transitions between QSs (4.1.12). The Fourier series expansion (4.1.10) of $X(t)$ has the frequencies $2q\omega$ only. It is clear that when $Q \approx 2q\omega$ multiphoton resonances between many QS harmonics arise, caused simultaneously by *all* harmonics of $X(t)$. The resonance condition may be finally written as

$$\langle W_n(t) \rangle = 2l\omega + \Delta, \quad l = 1, 2, 3, \ldots, \qquad (4.1.14)$$

where Δ is a detuning ($|\Delta| \ll \omega$). The possibility of multi-frequency resonances between QSs has been pointed out by *Anosov* [4.12].

In the resonance approximation, retaining in the set of equations in (4.1.9) the weakly oscillating terms only, we get a solution

$$c_{1,4} = \alpha_1 \exp(\pm it\Delta) + i\alpha_3(\Delta \mp \Omega)(2a)^{-1} \exp[i(\Omega \pm \Delta)t]$$

$$+ i\alpha_4(\Delta \pm \Omega)(2a)^{-1} \exp[-i(\Omega \mp \Delta)t], \quad c_3 = c_2 - i\alpha_1\Delta/a + 2\alpha_2,$$

$$c_2 = i\alpha_1\Delta/a - \alpha_2 + \alpha_3 \exp(i\Omega t) + \alpha_4 \exp(-i\Omega t). \qquad (4.1.15)$$

Here Ω is essentially the generalization of the Rabi frequency [4.13]:

$$\Omega = (\Delta^2 + 4a^2)^{1/2},$$

$$a(l) \equiv (Z\xi k^2\omega^2/16\varepsilon_0) \sum_{q=-\infty}^{+\infty} J_{q-l}(w)(g_{|2q-2|} - g_{|2q+2|}), \qquad (4.1.16)$$

where $w \equiv 3b_2\varepsilon_0/2\omega Z + Z\xi^2\omega g_2/12\varepsilon_0$; $J(w)$ are Bessel functions.

Four species of the initial conditions $c_j(0) = \delta_{pj}(j, \, p = 1, 2, 3, 4)$ give four sets of coefficients $\alpha_k^{(p)}$ in (4.1.15) and finally determine four orthonormal

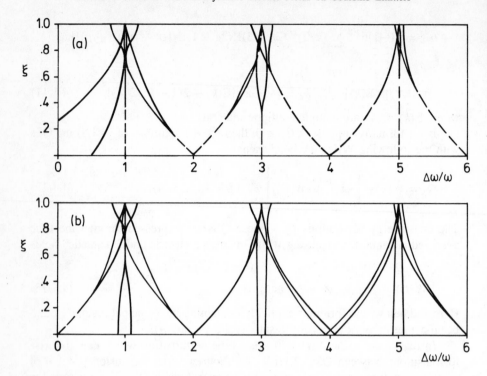

Fig. 4.2. The dependence of a L_α spectrum on the ellipticity degree ξ at $3E_0/Z\omega \equiv V = 5.5$. Intense spectral components are shown as bands (as double lines). Their position corresponds to the band median, their intensity is proportional to the band width: (**a**) x-polarization; (**b**) z-polarization

solutions of (4.1.5). As an example, in Fig. 4.2 it is shown how the L_α line spectrum depends on the ED ξ for $3E_0/(Z\omega) = 5.5$. In a vicinity of $\xi \approx 0.6$ the four-photon resonance ($l = 2$) occurs, which results in a drastic reconstruction of the spectrum (abrupt changes of intensities, appearance of new spectral components). Note that when $\xi \to 1$ the spectrum in the total space angle $I_x(\Delta\omega) + I_y(\Delta\omega) + I_z(\Delta\omega)$ turns into the spectrum obtained in [4.8,9].

From the condition $X^2(t) \ll 1$ it is not difficult to get the inequality determining the existence limit of the resonance effects:

$$\xi^2 \gg \frac{2\omega Z}{3n\varepsilon_0}. \tag{4.1.17}$$

Note that the values $a(l)$ may be found in the relatively simple form of (4.1.16) under the additional condition

$$\left| \frac{3n\varepsilon_0 b_4}{8\omega Z} + \frac{Z\xi^2 \omega g_4}{12n\varepsilon_0} \right| \ll 1. \tag{4.1.18}$$

It is interesting that in the case of "odd-photon-resonance" $\langle W_2(t) \rangle = (2l-1)\omega + \Delta$ the Rabi frequency is $\Omega = |\Delta|$ so that for zero detuning the QSs similar to (4.1.12) do not intermix at all. We point out that for the hydrogen atom in fields $F \perp E_0 \times \cos \omega t$ the analogous case ($\Omega = 0$) occurs for "even-photon resonance".

4.1.3 Elliptically Polarized Fields in the High-Frequency Limit

Using the general formalism of Sect. 2.3 the following results are obtained.

L_α *Line.* The four WFs $\psi_j(t)$ ($j = 1, 2, 3, 4$) of QSs are

$$\psi_j(t) = \exp[-\mathrm{i}(E_2 + \lambda_j)t] \sum_{k=1}^{4} \gamma_{kj} \varphi_j \exp[\mathrm{i}(n_2 - n_1)_j v \sin \omega t] \qquad (4.1.19)$$

with quasienergies $E_2 + \lambda_j$ where

$$E_n \equiv -Z^2/2n^2, \quad (n = 1, 2, \ldots); \quad \lambda_{1,2} = 0, \quad \lambda_3 = \kappa, \quad \lambda_4 = -\kappa;$$

$$\kappa \equiv \omega \xi v \mathrm{J}_1(v), \qquad v \equiv \frac{3nE_0}{2\omega Z(1 + \xi^2)^{1/2}}, \qquad (4.1.20)$$

$\gamma_{11} = \gamma_{22} = \gamma_{33} = \gamma_{44} = 1/2$, $\gamma_{12} = \gamma_{21} = \gamma_{34} = \gamma_{43} = -1/2$, $\gamma_{13} = \gamma_{31} = \gamma_{23} = \gamma_{32} = \gamma_{14} = \gamma_{41} = \gamma_{24} = \gamma_{42} = -\mathrm{i}/2$; $\varphi_1 \equiv |001\rangle$, $\varphi_2 \equiv |00-1\rangle$, $\varphi_3 \equiv |100\rangle$, $\varphi_4 \equiv |010\rangle$. Using the WFs (4.1.19) we find the spectrum of the L_α line:

$$I^{(e_x)}(\Delta\omega) = \delta(\Delta\omega - \kappa) + \delta(\Delta\omega + \kappa),$$

$$I^{(e_y)}(\Delta\omega) = 2\delta(\Delta\omega),$$

$$I^{(e_z)}(\Delta\omega) = \sum_{p=-\infty}^{+\infty} \{2\mathrm{J}_{2p+1}^2(v)\delta(\Delta\omega - (2p+1)\omega)$$

$$+ \mathrm{J}_{2p}^2(v)[\delta(\Delta\omega - 2p\omega + \kappa) + \delta(\Delta\omega - 2p\omega - \kappa)]. \qquad (4.1.21)$$

From (4.1.21) it follows that at $\xi = 0$ the spectrum coincides with the L_α spectrum found by *Blochinzew* [4.14]. With an increase of ξ a symmetrical splitting of the central component and of each even satellite (of lateral components) occurs. Odd satellites of lateral components do not split.

Arbitrary hydrogen-like line. The *dynamical* problem of calculation of quasienergies $E_n + \lambda$ in an elliptically polarized QEF may be formally reduced to some *static* Stark effect in some *effective* field F_{eff}:

$$\lambda = (3n/2)(n_1 - n_2)F_{\mathrm{eff}}, \qquad F_{\mathrm{eff}} \equiv \xi \varepsilon_0 \mathrm{J}_1(3n\varepsilon_0/(2\omega)). \qquad (4.1.22)$$

The following WFs of QSs correspond to these quasienergies $E_n + \lambda$:

$$\psi_\lambda(t) = \exp[-\mathrm{i}(E_n + \lambda)t] \sum_{k,s} (-\mathrm{i})^s d_k^{(s)} \exp[-\mathrm{i}(\langle \varphi_k^{(s)} | z | \varphi_k^{(s)} \rangle \varepsilon_0/\omega) \sin \omega t].$$

$$(4.1.23)$$

Here $s = n - n_1 + n_2 - 1$, $\varphi_k^{(s)}$ are WFs in parabolic coordinates with a quantization axis orthogonal to the field plane; an index k numerates states with fixed s. Coefficients $d_k^{(s)}$ have the following form:

$$d_k^{(s)} = \langle \varphi_k^{(s)} | \exp(-\mathrm{i}l_y \pi/2) | \varphi_{n_1 n_2 m} \rangle. \qquad (4.1.24)$$

The region of validity of the obtained results is determined by the inequality $\omega \gg |\lambda|$.

4.1.4 Measurements of Elliptically Polarized Field Parameters

The nonlinear optical effect considered above is interesting both in theory and in applied problems. It enables one to obtain, for strong fields $3nE_0/2\omega Z \equiv V \gg 1$, the analytical expression for quasienergies $\varepsilon(\xi, V)$ practically in the entire range of Ellipticity Degree (ED) values $0 \leqslant \xi \leqslant 1$. Indeed in the range $\xi \lesssim V^{-1/2}$, the quasienergies may be determined analytically from perturbation theory. In particular, for $n = 2$ we have

$$\varepsilon_\pm/\omega = \pm \xi v \mathrm{J}_1(v), \quad v \equiv V(1 + \xi^2)^{-1/2}. \tag{4.1.25}$$

And in the range $V^{-1/2} \ll \xi \leqslant 1$, see (4.1.17) the quasienergies may be determined analytically by the resonance approximation. In particular, for $n = 2$

$$\varepsilon_\pm/\omega = \pm(-1)^l(\Omega/\omega - 1), \tag{4.1.26}$$

where l is the resonance number in (4.1.14). Note that for $\xi = 1$ the quasienergies from (4.1.26) correspond to the exact solutions for circular polarization [4.8, 9].

The functions $\varepsilon(\xi)$ $(n = 2)$ for various values of the parameters v and V are shown in Fig. 4.3. Since the quasienergies define the observed positions of spectral components (say, for L_α lines) in principle it should be possible to measure the ED directly in plasmas. It is enough to measure beforehand the total amplitude E_0 (for example, by the helium satellites method, which in many cases is nearly independent of ED) and hence the parameter V. Then by comparing the observed positions of quasienergy harmonics with the theoretical function $\varepsilon(\xi)$, it is not difficult to determine ξ.

This method of ED measurements could be useful in the study of microwave heating in tokamaks. Indeed, the ED of strong microwave radiation controls both the efficiency of heating and the limiting value of plasma density allowing effective heating [4.15]. The ED in plasma ξ_{pl} may differ from the known ED in the incident wave ξ_{vac} for various reasons, for example, because of reflections from the chamber walls.

Thus using the L_α line spectra of atoms H I, D I or ions He II, Li III, Be IV (injected into plasma) one can determine the ED ξ_{pl} and hence the efficiency of microwave heating in the range of temperatures $T \sim 10^2 - 10^3$ eV and even $T \gtrsim 1$ keV.

4.1.5 Analytical Investigation of Two-Dimensional Multimode QEFs[1]

We want to find the splitting of the hydrogen SL L_α in a two-dimensional multimode QEF of the form

---

[1] The results of this subsection have been jointly obtained with V.P. Gavrilenko.

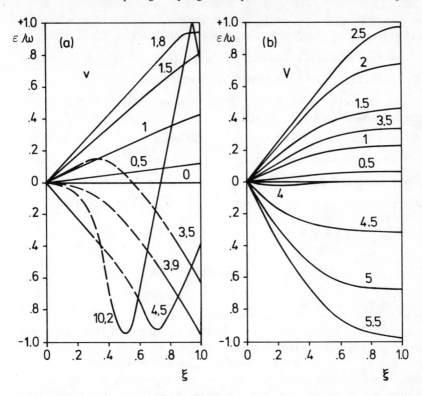

Fig. 4.3. The dependence of reduced quasienergies ε/ω on ellipticity degree ξ for $n = 2$ at different values of v (**a**) and V (**b**). *Dashed* parts of curves are interpolations between analytical calculations

$$E(t) = E_x(t)e_x + E_z(t)e_z,$$

$$E_x(t) = \sum_k E_k \cos(\omega t + \varphi_k) = \rho_x \cos(\omega t + \alpha_x),$$

$$E_z(t) = \sum_p E_p \cos(\omega t + \varphi_p) = \rho_z \cos(\omega t + \alpha_z), \qquad (4.1.27)$$

where the phases α_x, α_z are uniformly distributed in the interval $(0, 2\pi)$ and the amplitudes ρ_x, ρ_z have Rayleigh distributions $W(\rho_x)$, $W(\rho_z)$, see (3.2.7). For simplicity we consider a situation in which $\rho_x \ll \rho_z$.

We try a solution of the Schrödinger equation

$$i \, \partial \psi / \partial t = [H_a + x E_x(t) + z E_z(t)]\psi \qquad (4.1.28)$$

in the form

$$\psi = C_1 \varphi_1 + C_2 \varphi_2 + C_3 \exp[-iv_z \sin(\omega t + \alpha_z)]$$
$$+ C_4 \exp[iv_z \sin(\omega t + \alpha_z)], \qquad (4.1.29)$$

where $v_z \equiv 3\rho_z/\omega$; φ_1, φ_2, φ_3, φ_4 are parabolic WFs of the level with $n = 2$ defined according to (4.1.20). Substituting (4.1.29) into (4.1.28) we obtain

$$\dot{C}_{1,2} = (3/2)iE_x(t)\{C_3 \exp[-iv_z \sin(\omega t + \alpha_z)] + C_4 \exp(iv_z \sin(\omega t + \alpha_z)]\},$$

$$\dot{C}_{3,4} = (3/2)iE_x(t)(C_2 + C_3) \exp[\pm iv_z \sin(\omega t + \alpha_z)]. \tag{4.1.30}$$

Expanding exponents in (4.1.30) in Bessel functions and using the so-called "averaging principle" /4.16/ we find

$$\dot{C}_{1,2} = (K/2)(C_4 - C_3), \qquad \dot{C}_{3,4} = \pm(K/2)(C_1 + C_2);$$

$$K \equiv -3\rho_x J_1(v_z) \sin(\alpha_z - \alpha_x). \tag{4.1.31}$$

The spectrum of L_α at a fixed phase difference $\alpha_z - \alpha_x$ has the form

$$I^{(e_x)}(\Delta\omega) = \delta(\Delta\omega - K) + \delta(\Delta\omega + K), \qquad I^{(e_y)}(\Delta\omega) = 2\delta(\Delta\omega),$$

$$I^{(e_z)}(\Delta\omega) = \sum_{p=-\infty}^{+\infty} \{2J_{2p+1}^2(v_z)\delta(\Delta\omega - (2p+1)\omega)$$

$$+ J_{2p}^2(v_z)[\delta(\Delta\omega - 2p\omega + K) + \delta(\Delta\omega - 2p\omega - K)],$$

$$v_z \equiv 3\rho_z/\omega. \tag{4.1.32}$$

From (4.1.32) it is seen that the multidimensionality of a QEF leads to the appearance of *nonzero quasienergies* $\pm K$ in the L_α spectrum. This result means that the oversimplified theoretical approach to the same problem, which has been suggested in [4.17] and has led to zero quasienergies, is not valid.

Since the quasienergies depend upon a phase difference $\alpha_z - \alpha_x$ the averaging over $\alpha_z - \alpha_x$ leads to the transformation of a splitting into a broadening. So, e.g. for one of the δ-functions in the spectrum of $I^{(e_x)}(\Delta\omega)$ from (4.1.32) we have

$$\langle I^{(e_x)}(\Delta\omega)\rangle = (2\pi)^{-1} \int_0^{2\pi} d(\alpha_z - \alpha_x)\delta(\Delta\omega - 3\rho_x J_1(3\rho_z/\omega) \sin(\alpha_z - \alpha_x))$$

$$= \pi^{-1}\{[3\rho_x J_1(3\rho_z/\omega)]^2 - (\Delta\omega)^2\}^{-1/2}. \tag{4.1.33}$$

Further averaging of the spectrum over the Rayleigh distributions $W(\rho_x)$, $W(\rho_z)$ of amplitudes ρ_x, ρ_z may be performed by numerical methods.

The results obtained for the two-dimensional case are true in the general case. The point is that the problem involving a three-dimensional multimode QEF may be always reduced to a two-dimensional problem. Indeed such a field may be presented in the form

$$E(t) = e_x\rho_x \cos\omega t + e_y\rho_y \cos(\omega t + \alpha_y) + e_z\rho_z \cos(\omega t + \alpha_z). \tag{4.1.34}$$

It can be easily checked that at any time t the vector $E(t)$ belongs to the plane of the following stationary vectors:

$$E_1 = -e_y\rho_y \sin\alpha_y - e_z\rho_z \sin\alpha_z,$$

$$E_2 = e_x\rho_x \sin\alpha_y + e_z\rho_z \sin(\alpha_y - \alpha_z) \tag{4.1.35}$$

and thus is really two-dimensional.

4.2 Joint Action of QEF and Quasistatic EF on Hydrogen-like Spectral Lines

4.2.1 Dynamic Resonance

We want to investigate the splitting of hydrogen-like SLs in an EF

$$E(t) = F + E_D(t), \qquad E_D(t) = E_0 \cos \omega t \qquad (4.2.1)$$

paying special attention to nonadiabatic effects. One can expect the nonadiabatic effects to be maximized for $F \perp E_0$ [4.18]. Therefore we analyze this case first.

We choose the z axis of the fixed coordinate system along E_0 and the x axis along F. We change over to a rotating frame $x'y'z'$, whose z' axis is directed at each instant of time along the resultant field $E(t)$ and makes an angle $\varphi(t)$ with the z axis.

The Hamiltonian of the atom in the $x'y'z'$ system is of the form

$$H = H_a + V_1(t) + V_2(t), \quad V_1(t) = zE(t), \quad V_2(t) = l_y \dot\varphi(t),$$

$$E(t) \equiv |E(t)|, \qquad \dot\varphi = \omega F E_0 (\sin \omega t)/E^2(t). \qquad (4.2.2)$$

Here H_a is the Hamiltonian of the isolated atom in the xyz system; l_y is the projection of the angular momentum on the y axis. The dc component $V_1(t)$ of the perturbation splits the states with equal n into $2n - 1$ sublevels with a spacing between them of

$$\omega_{\bar E} = 3\pi^{-1} n (F^2 + E_0^2)^{1/2} \mathbb{E}(k), \qquad (4.2.3)$$

where $k \equiv E_0 (F^2 + E_0^2)^{-1/2}$, and $\mathbb{E}(k)$ is the complete elliptic integral of the second kind in its normal form.

If we put $V_2(t) \equiv 0$ in the Hamiltonian (4.2.2), then the solutions of the Schrödinger equation can be chosen to be QSs with WFs $\psi_\alpha(t)$:

$$\psi_\alpha(t) = \exp[-i(n_1 - n_2)\omega_{\bar E} t]\Phi_\alpha(t),$$

$$\Phi_\alpha(t) = \exp\left\{ i \left[\omega_{\bar E} t - (3/2)n \int_0^t dt'\, E(t') \right] (n_1 - n_2) \right\} \psi_\alpha(r) \qquad (4.2.4)$$

$$= \sum_{s=-\infty}^{+\infty} \Phi_\alpha^s(r) \exp(2is\omega t).$$

Here $\alpha \equiv (n_1, n_2, m)$ are the parabolic quantum numbers; $\Phi_\alpha^s(r)$ are Fourier coefficients. We thus obtain the QSs with quasienergies separated by

$$Q = \omega_{\bar E} + 2u\omega \quad (u = 0, \pm 1, \pm 2, \ldots).$$

The "magnetic" perturbation $V_2(t)$ has the following Fourier expansion:

$$V_2(t) = l \sum_{v=1}^{\infty} b_v \sin(2v - 1)\omega t,$$

$$b_v = 2(-1)^{v+1}[(1 + F^2 E_0^{-2})^{1/2} - F E_0^{-1}]^{2v-1}\omega. \qquad (4.2.5)$$

It is clear that at $Q = (2v - 1)\omega$ multiphoton resonance sets in between many QS harmonics and is due simultaneously to *all* the harmonics of $V_2(t)$ (*multi-frequency* resonance). The final form of the resonance condition is

$$\omega_{\bar{E}} = (2l - 1)\omega, \quad l = 1, 2, 3, \dots . \tag{4.2.6}$$

The physical meaning of this is most evident in the case $E_0 \gg F$, when the dc component of the fields is $\bar{E} \equiv \langle |E(t)| \rangle \approx 2E_0/\pi$. Condition (4.2.6) then means that an odd harmonic of the dynamic field frequency coincides with the splitting $\omega_{\bar{E}} \approx (3/2)n(2E_0/\pi)$ arising from the same *dynamic* field. That is why this phenomenon is called dynamic resonance. We note that although the static field does not enter in the resonance condition at $E_0 \gg F$, it determines the amplitudes b_v of the harmonics of the magnetic perturbation $V_2(t)$, see (4.2.5), and the produced QS splitting.

4.2.2 Hydrogen-like Lines at a Multiquantum Dynamic Resonance and Away from the Resonance

Consider a general case in which F and $E_D(t)$ fields are not necessarily orthogonal. We choose a coordinate system with the origin at the nucleus of the hydrogen atom, the z axis running along the field F, and the x axis in the plane defined by the vectors F and E_0 (Fig. 4.4). The WF of the hydrogen atom, $\psi(t)$, then satisfies the Schrödinger equation

$$i\,\partial\psi/\partial t = H\psi, \quad H = H_a + z[F + E_{0z}\cos(\omega t + \varphi_0)] + x E_{0x}\cos(\omega t + \varphi_0). \tag{4.2.7}$$

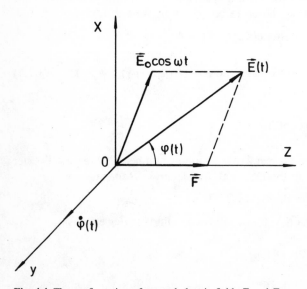

Fig. 4.4. The configuration of crossed electric fields F and $E_0\cos\omega t$

Here H_a is the Hamiltonian of the unperturbed atom, which has a discrete spectrum $\mathcal{E}_n$; E_{0x} and E_{0z} are the projections of E_0 onto the x and z axes, and the operators x and z represent the x and z projections of the dipole moment r. We transform to the new WF $\Phi(t)$, defined by

$$\Phi(t) = \exp(i\varphi(t)l_y)\psi(t), \tag{4.2.8}$$

where $\varphi(t)$ is the angle between the z axis and the resultant field $E(t) = F + E_0 \cos \omega t$, and the operator l_y represents the y projection of the angular momentum. Substituting (4.2.8) into (4.2.7), we find the following equation for the wave function $\Phi(t)$:

$$i \partial \Phi/\partial t = [H_a + V_1(t) + V_2(t)]\Phi,$$

$$V_1(t) = zE(t) = z(F^2 + E_0^2 \cos^2 \omega t + 2FE_{0z} \cos \omega t)^{1/2},$$

$$V_2(t) = -l_y \dot{\varphi}(t), \quad \dot{\varphi}(t) = -FE_{0x}\omega(\sin \omega t)/E^2(t). \tag{4.2.9}$$

Expansions of the functions $\Phi(t)$ and $E(t)$ in Fourier series are

$$\dot{\varphi}(t) = (i/2) \sum_{p=-\infty}^{+\infty} (\text{sign } p)b_{|p|} \exp(ip\omega t),$$

$$E(t) = \varepsilon_0/2 + \sum_{j=1}^{+\infty} \varepsilon_j \cos j\omega t. \tag{4.2.10}$$

Here we have taken sign $(0) = 0$. The coefficients b_q and ε_j are given explicitly in Appendix C.

We write the unknown function $\Phi(t)$ in the form

$$\Phi(t) = \exp(-i\lambda t) \exp\left[-iH_a t - i \int_0^t V_1(t')\,dt'\right] \varphi^{(\lambda)},$$

$$\varphi^{(\lambda)} = \sum_{k'=1}^{n^2} c_{k'}^{(\lambda)} \varphi_{k'}, \tag{4.2.11}$$

where $\Phi_{k'} \equiv \Phi_{n_1 n_2 m}$ is the wave function in parabolic coordinates, with the z axis as the quantization axis, which belongs to the level $\mathcal{E}_n$.

Let us examine the case of a multiquantum resonance, the condition for which is

$$3n\varepsilon_0/4 = q\omega, \quad q = 1, 2, 3, \ldots \tag{4.2.12}$$

A resonance is possible for an arbitrary angle between the vectors F and $E_D(t)$ not only at odd but also at even values of q [in contrast to the case with $F \perp E_D(t)$].

To calculate the resonance we substitute the wave function $\Phi(t)$ from (4.2.11) into (4.2.9):

$$\lambda c_k^{(\lambda)} = \sum_{k'=1}^{n^2} c_{k'}^{(\lambda)} V_{kk'}(t), \quad k = 1, 2, \ldots, n^2, \tag{4.2.13}$$

where

$$\Delta z(k, k') = z_{kk} - z_{k'k'}. \tag{4.2.14}$$

It follows from (4.2.10) and from the condition for a q-quantum resonance, (4.2.12), that the nonzero matrix elements $V_{kk'}(t)$ given by (4.2.14) contain a constant component $(V_q)_{kk'}$:

$$(\bar{V}_q)_{kk'} = R_q^{(0)} x_{kk'}/(3n),$$

$$R_q^{(0)} = \sum_{r,s=-\infty}^{+\infty} J_r(2q\varepsilon_1/\varepsilon_0) J_s(q\varepsilon_2/\varepsilon_0)[\mathrm{sign}(q+r+2s)]b_{|q+r+2s|}, \tag{4.2.15}$$

where, as in (4.2.10), sign $(0) = 0$, and $J_m(x)$ are Bessel functions. For simplicity, we have restricted the derivation of (4.2.15) to the case

$$|2q\varepsilon_p/p\varepsilon_0| \ll 1, \quad p = 3, 4, \dots, \tag{4.2.16}$$

and made use of the following relation between matrix elements:

$$(l_y)_{kk'} = -2ix_{kk'}(3n)^{-1} \mathrm{sign}[\Delta z(k, k')]. \tag{4.2.17}$$

We will restrict the discussion to the constant components in solving system (4.2.13) in the resonance approximation. Substituting (4.2.15) into (4.2.13), we find equations describing the static Stark effect in an effective electric field $R_q^{(0)}/3n$, which is directed along the x axis. The eigenvalues λ are therefore

$$\lambda(q) = (n_1 - n_2) R_q^{(0)}/2. \tag{4.2.18}$$

Corresponding to these eigenvalues[2] are the WFs $\varphi^{(\lambda)}$:

$$\varphi^{(\lambda)} = \exp(-il_y\pi/2)\varphi_{n_1n_2m}. \tag{4.2.19}$$

Using (4.2.8, 11, 19), we find the WFs of QSs $\psi_{n_1n_2m}^{(q)}(t)$, which corresponds to quasienergies

$$\psi_{n_1n_2m}^{(q)}(t) = \exp[-i\mathcal{E}_n t - i(n_1 - n_2)R_q^{(0)}t/2]\exp[-i\varphi(t)l_y]$$

$$\times \exp\left[-iz\int_0^t dt' \, E(t')\right]\exp(-i\pi l_y/2)\varphi_{n_1n_2m}. \tag{4.2.20}$$

Far from the resonance, $|\Delta| = |3n\varepsilon_0/4 - q\omega| \gg |R_q^{(0)}|$, the solutions of the Schrödinger equation (4.2.7) are the following WFs:

$$\psi_{n_1n_2m}(t) = \exp[-i\varphi(t)l_y]\Phi_{n_1n_2m}(t),$$

$$\Phi_{n_1n_2m}(t) = \exp\left[-i\mathcal{E}_n t - (3i/2)n(n_1 - n_2)\int_0^t dt' \, E(t')\right]\varphi_{n_1n_2m}. \tag{4.2.21}$$

[2] In the particular case of a single-quantum resonance ($q = 1$, $E_0 \ll F$), the values of λ are $\lambda = 3n(n_1 - n_2)E_0/4$.

The range of applicability of (4.2.20) and (4.2.21) is determined by the condition

$$P = \frac{1}{36} \frac{E_{0x}^2 \omega^2 (F^2 + 3E_0^2)^4}{F^4} \cdot \frac{F^4 - 3F^2 E_0^2 + 9E_0^4 + 9F^2 E_{0x}^2}{(F^4 + 9E_0^2 E_{0x}^2 + 6F^2 E_{0x}^2)^3} \ll 1,$$

(4.2.22)

under which the perturbation $V_2(t)$ in (4.2.9) need be considered only near the resonance (4.2.12).

From (4.2.20, 21) we can find the analytic spectrum of any hydrogen line for any static and dynamic fields satisfying condition (4.2.22), except the case of intermediate deviations from the resonant frequency, $|\Delta| \sim |R_q^{(0)}|$ for the initial or final state. In the latter case, the frequencies of the additional splitting for an arbitrary level n are $\lambda \approx (n_1 - n_2)[\Delta^2 + (R_q^{(0)}/2)^2]^{1/2}$, but we will not reproduce here the lengthy expressions for the WFs.

Let us consider the situation in which there is no resonance for either the initial level a or the final level b. A calculation of the spectrum of the hydrogen line, $I_{ab}^{(e)}(\Delta\omega)$, in the polarizations e_x, e_y, e_z (the unit vectors along the x, y, and z axes) then yields

$$I_{ab}^{(e)}(\Delta\omega) = \sum_{\alpha \in a, \beta \in b} I_{\alpha\beta}^{(e)}(\Delta\omega),$$

$$I_{\alpha\beta}^{e}(\Delta\omega) = \sum_{k=-\infty}^{+\infty} A_{k,\alpha\beta}^{(e)} \delta(\Delta\omega - (z_{\alpha\alpha} - z_{\beta\beta})\varepsilon_0/2 - k\omega),$$

$$A_{k,\alpha\beta}^{(e_x)} = \left| \sum_{p,r=-\infty}^{+\infty} Q_{k+p+r}^{(\alpha)} Q_p^{(\beta)} (z_{\alpha\beta} d_r^{(s)} + x_{\alpha\beta} d_r^{(c)}) \right|^2,$$

$$A_{k,\alpha\beta}^{(e_y)} = \left| \sum_{p=-\infty}^{+\infty} Q_{k+p}^{(\alpha)} Q_p^{(\beta)} y_{\alpha\beta} \right|^2,$$

$$A_{k,\alpha\beta}^{(e_z)} = \left| \sum_{p,r=-\infty}^{+\infty} Q_{k+p+r}^{(\alpha)} Q_p^{(\beta)} (z_{\alpha\beta} d_r^{(c)} - x_{\alpha\beta} d_r^{(s)}) \right|^2,$$

(4.2.23)

where $\xi_{\alpha\beta} = \langle \varphi(n_1 n_2 m)_\alpha | \xi | \varphi(n_1 n_2 m)_\beta \rangle$; $\xi = x, y, z$ and $d_r^{(s)}$, $d_r^{(c)}$, and $Q_k^{(\mu)}$ are given by

$$d_r^{(s)} \approx \text{Im}\{\kappa_r\}, \quad d_r^{(c)} \approx \text{Re}\{\kappa_r\},$$

$$\kappa_{-r} = \kappa_r \equiv \exp[ib_2/(2\omega) + \pi r/2]$$

$$\times \sum_{p,s=-\infty}^{+\infty} \exp(-i\pi p/2) J_{r-2p-3s}(b_1/\omega) J_p(b_2/2\omega) J_s(-b_3/3\omega),$$

$$Q_k^{(\mu)} \approx \sum_{s=-\infty}^{+\infty} J_{k-2s}(z_{\mu\mu}\varepsilon_1/\omega) J_s(z_{\mu\mu}\varepsilon_2/2\omega), \quad z_{\mu\mu} = (3/2)n_\mu(n_1 - n_2)_\mu.$$

(4.2.24)

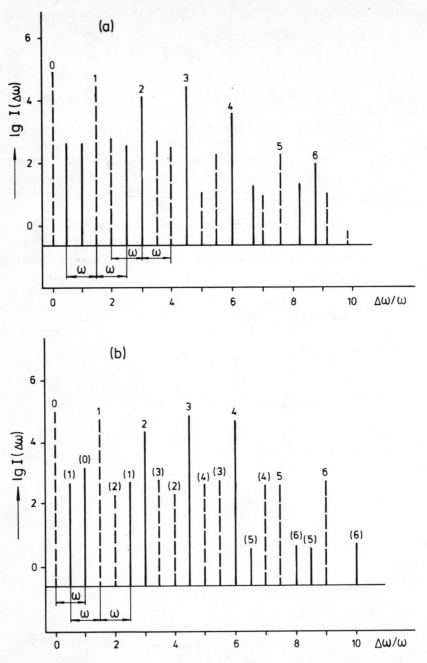

Fig. 4.5. Comparison of theoretical profiles of the H_α SL in orthogonal fields F and $E_0 \cos \omega t$ at $F/\omega = 1$, $E_0/\omega = 2^{3/2}/15$: (**a**) analytical results; (**b**) numerical results from [4.18]. *Solid lines* – polarizer axis parallel to F, *dashed lines* – polarizer axis parallel to E_0. Satellites marked in (**b**) by numbers in parentheses are $\pm \omega$ away from the components marked by the same numbers without parentheses (these components exist also at $E_0 = 0$)

In (4.2.23), the origin of the frequency scale $\Delta\omega$ is at the center of the SL corresponding to the transition $a \to b$. The spectra found for the H_α line in [4.18] by numerical calculation agree fairly well with the results calculated from (4.2.23) (Fig. 4.5).

We turn now to the case in which a resonance is missing from only one of the two levels between which a radiative transition occurs; for the other level, an exact q-quantum resonance as in (4.2.12) occurs. Let us assume, for example, that resonance (4.2.12) holds for the initial level a. The WFs of the states α are then given by (4.2.20), and those of the states β by (4.2.21).

Going through the calculation of the spectrum as in the case in which there is no resonance for either level, we easily find

$$I_{\alpha\beta}^{(e)}(\Delta\omega) = \sum_{k=-\infty}^{+\infty} B_{k,\alpha\beta}^{(e)}\delta(\Delta\omega - (n_1 - n_2)_\alpha R_q^{(0)}/2 + z_{\beta\beta}\varepsilon_0/2 - k\omega),$$

$$B_{k,\alpha\beta}^{(e_x)} = \left| \sum_{\alpha' \in a} \sum_{p,r=-\infty}^{+\infty} C_{\alpha\alpha'} Q_{p+r+k-q(n_1-n_2)_{\alpha'}}^{(\alpha')} Q_p^{(\beta)} \left(z_{\alpha'\beta}d_r^{(s)} + X_{\alpha'\beta}d_r^{(c)} \right) \right|^2,$$

$$B_{k,\alpha\beta}^{(e_y)} = \left| \sum_{\alpha' \in a} \sum_{p=-\infty}^{+\infty} C_{\alpha\alpha'} Q_{p+k-q(n_1-n_2)_{\alpha'}}^{(\alpha')} Q_p^{(\beta)} y_{\alpha'\beta} \right|^2,$$

$$B_{k,\alpha\beta}^{(e_z)} = \left| \sum_{\alpha' \in a} \sum_{p,r=-\infty}^{+\infty} C_{\alpha\alpha'} \times Q_{p+r+k-q(n_1-n_2)_{\alpha'}}^{(\alpha')} \right.$$

$$\left. Q_p^{(\beta)} \left(z_{\alpha'\beta}d_r^{(c)} - x_{\alpha'\beta}d_r^{(s)} \right) \right|^2 ;$$

$$C_{\alpha\alpha'} \equiv \langle \varphi(n_1 n_2 m)_\alpha | \exp(i\pi l_y/2) | \varphi(n_1 n_2 m)_{\alpha'} \rangle. \qquad (4.2.25)$$

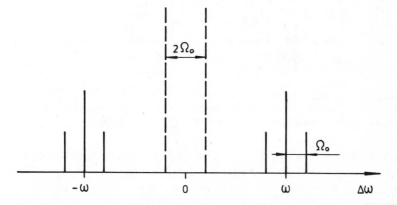

Fig. 4.6. The spectrum of the L_α line in orthogonal fields F and $E_0 \cos\omega t$ under the single-quantum resonance in the $n = 2$ level. The direction of the observation is assumed to be perpendicular to F and E_0, positions of the transmission axis of the polarizer are parallel to F (*solid lines*) or parallel to E (*dashed lines*). The splitting is equal to $\Omega_0 = 3E_0/2$

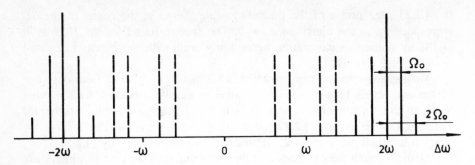

Fig. 4.7. Same as in Fig. 4.6 but for the L_β line under the single-quantum resonance in the $n = 3$ level ($\Omega_0 = 9E_0/4$)

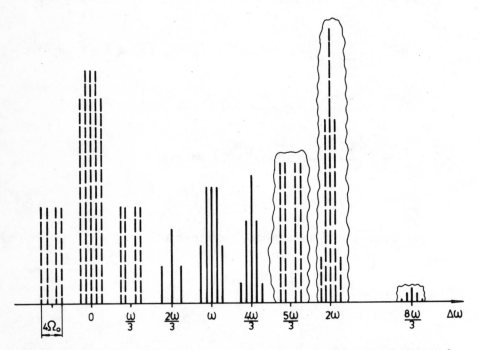

Fig. 4.8. Same as in Fig. 4.7 but for the H_α line. The entire spectrum is shown only at $\Delta\omega \geqslant 0$, since the spectrum is symmetric with respect to the unperturbed frequency $\Delta\omega = 0$. The intensities of the components inside the *wavy lines* are shown enlarged by a factor of 200

Figures 4.6–8 show the spectra of the L_α, L_β, and H_α lines at an exact single-quantum resonance ($q = 1$, $E_0 \ll F$) for the upper level.

In the case of intermediate deviations from the resonant frequency, $|\Delta| \sim |R_q^{(0)}|$, the emission spectrum can be determined by an interpolation of the spectra found at small and large deviations from resonance. We turn now to a detailed study of the spectrum, with allowance for the deviation from the resonant frequency, in the particular case of the L_α line of hydrogen.

4.2.3 The L_α Spectral Line with Detuning from Resonance

In a real plasma, different hydrogen atoms are generally acted upon by different quasistatic fields F, which induce arbitrary deviations from the resonant frequency. Let us examine the effect of such a frequency deviation in the particular case of the L_α hydrogen line. We retain the assumptions used in Sect. 4.2.2 regarding the static and dynamic fields, but we replace condition (4.2.12) for $n = 2$ by the resonance condition

$$3\varepsilon_0/2 = q\omega + \Delta, \quad q = 0, 1, 2, \dots \tag{4.2.26}$$

(the value $q = 0$ corresponds to the static Stark effect: $E_0 = 0$).

We introduce the following notation for the Stark states with $n = 2$:

$$\varphi_{001} \equiv \varphi_1, \quad \varphi_{00-1} \equiv \varphi_2, \quad \varphi_{100} \equiv \varphi_3, \quad \varphi_{010} \equiv \varphi_4. \tag{4.2.27}$$

As in Sect. 4.2.2, we seek a solution of (4.2.7) in the form of (4.2.11), but we write the WF $\varphi^{(\lambda)}$ as

$$\varphi^{(\lambda)} = \sum_{k=1}^{4} C_k^{(\lambda)} [\exp(\mathrm{i}(n_1 - n_2)_k t\Delta)]\varphi_k. \tag{4.2.28}$$

In the resonant approximation for $n = 2$, we can easily find a generalization of (4.2.15) to the case $\Delta \neq 0$, $|\Delta| \ll \omega$:

$$(\bar{V}_q)_{kk'} = R_q^{(\Delta)} x_{kk'}/6, \quad R_q^{(\Delta)} = \sum_{r,s=-\infty}^{+\infty} \mathrm{J}_r\left(\frac{2\varepsilon_1}{\varepsilon_0}\left(q + \frac{\Delta}{\omega}\right)\right).$$

$$\mathrm{J}_s\left(\frac{\varepsilon_2}{\varepsilon_0}\left(q + \frac{\Delta}{\omega}\right)\right) b_{|q+r+2s|} \, \mathrm{sign}(q + r + 2s). \tag{4.2.29}$$

Using (4.2.29) in the resonant approximation with $|\Delta| \ll \omega$, we replace the system (4.2.13) by

$$4\lambda C_1^{(\lambda)} + R_q^{(\Delta)} C_3^{(\lambda)} + R_q^{(\Delta)} C_4^{(\lambda)} = 0, \quad \lambda(C_1^{(\lambda)} - C_2^{(\lambda)}) = 0,$$

$$R_q^{(\Delta)} C_1^{(\lambda)} + R_q^{(\Delta)} C_2^{(\lambda)} + 4(\lambda - \Delta)C_3^{(\lambda)} = 0,$$

$$R_q^{(\Delta)} C_1^{(\lambda)} + R_q^{(\Delta)} C_2^{(\lambda)} + 4(\lambda + \Delta)C_4^{(\lambda)} = 0, \tag{4.2.30}$$

where we have made use of the fact that at $n = 2$ all the nonvanishing matrix elements are equal to $-3/2$. Equating the determinant of system (4.2.30) to zero, we find

$$\lambda_{1,2} = 0, \quad \lambda_3 = \Omega_\Delta(q) = [\Delta^2 + (R_q^{(\Delta)}/2)^2]^{1/2}, \quad \lambda_4 = -\Omega_\Delta(q). \tag{4.2.31}$$

Once we have found the coefficients $C_k^{(\lambda)}$ from the system in (4.2.30), we can easily use (4.2.11, 28) to find expressions for the WFs $\Phi_j(t)$ corresponding to the values λ_j $(j = 1, 2, 3, 4)$ in (4.2.31):

$$\Phi_j(t) = \exp[-i(\mathcal{E}_2 + \lambda_j)t][\alpha_{1j}\varphi_1 + \alpha_{2j}\varphi_2 + \alpha_{3j}\exp(-iq\omega t)\sum_{k=-\infty}^{+\infty}Q_k$$

$$\times \exp(-ik\omega t) + \sum_{k=-\infty}^{+\infty}Q_k\exp(-ik\omega t)\varphi_3$$

$$+ \alpha_{4j}\exp(iq\omega t)\sum_{k=-\infty}^{+\infty}Q_k\exp(ik\omega t)\varphi_4], \quad j = 1, 2, 3, 4,$$

$$(4.2.32)$$

where

$$\alpha_{11} = \alpha_{22} = \alpha_{33} = \alpha_{44} = (\Delta + \Omega_\Delta)/(2\Omega_\Delta),$$
$$\alpha_{12} = \alpha_{21} = -\alpha_{34} = -\alpha_{43} = (\Delta - \Omega_\Delta)/(2\Omega_\Delta),$$
$$-\alpha_{13} = \alpha_{31} = \alpha_{14} = -\alpha_{41} = -\alpha_{23} = \alpha_{32} = \alpha_{24} = -\alpha_{42} = R_q^{(\Delta)}/(4\Omega_\Delta),$$

$$(4.2.33)$$

and the Q_k are given by

$$Q_k \equiv Q_k^{(100)} \approx \sum_{s=-\infty}^{+\infty} J_{k-2s}\left(\frac{2\varepsilon_1}{\varepsilon_0}\left(q + \frac{\Delta}{\omega}\right)\right) J_s\left(\frac{\varepsilon_2}{\varepsilon_0}\left(q + \frac{\Delta}{\omega}\right)\right). \quad (4.2.34)$$

Going through the calculation of the spectrum as in Sect. 4.2.2, we find

$$I^{(e)}(\Delta\omega) = \sum_{j=1}^{4} I_j^{(e)}(\Delta\omega), \qquad I^{(e_y)}(\Delta\omega) = 2\delta(\Delta\omega),$$

$$I_j^{(e_x)}(\Delta\omega) = \sum_{p=-\infty}^{+\infty}\delta(\Delta\omega - \lambda_j + p\omega)\Big|(\alpha_{1j} + \alpha_{2j})d_p^{(c)}$$

$$-\alpha_{3j}\sum_{r=-\infty}^{+\infty}Q_{-p+r-q}d_r^{(s)} + \alpha_{4j}\sum_{r=-\infty}^{+\infty}Q_{p-r-q}d_r^{(s)}\Big|^2,$$

$$I_j^{(e_z)}(\Delta\omega) = \sum_{p=-\infty}^{+\infty}\delta(\Delta\omega - \lambda_j + p\omega)\Big|(\alpha_{1j} + \alpha_{2j})d_p^{(s)}$$

$$+\alpha_{3j}\sum_{r=-\infty}^{+\infty}Q_{-p+r-q}d_r^{(c)} - \alpha_{4j}\sum_{r=-\infty}^{+\infty}Q_{p-r-q}d_r^{(c)}\Big|^2, \quad (4.2.35)$$

where q is the index of the resonance.

In the case of a single-quantum resonance $(q = 1, E_0 \ll F)$,

$$3F = \omega + \Delta \qquad (4.2.36)$$

the expressions for $I^{(e_x)}(\Delta\omega)$ and $I^{(e_z)}(\Delta\omega)$ simplify:

$$I^{(e_x)}(\Delta\omega) = 2(\Delta/\Omega_\Delta)^2\delta(\Delta\omega)$$
$$+ [3E_{0x}/(2\Omega_\Delta)]^2[\delta(\Delta\omega - \Omega_\Delta) + \delta(\Delta\omega + \Omega_\Delta)],$$

$$I^{(e_z)}(\Delta\omega) = \frac{9E_{0x}^2}{8\Omega_\Delta^2}\left[\delta(\Delta\omega - \omega) + \delta(\Delta\omega + \omega) + \left(\frac{9E_{0x}^2}{8\Omega_\Delta(\Omega_\Delta - \Delta)}\right)^2\right.$$

$$\times [\delta(\Delta\omega - \Omega_\Delta - \omega) + \delta(\Delta\omega + \Omega_\Delta + \omega)]$$

$$\left.+ \left(\frac{9E_{0x}^2}{8\Omega_\Delta(\Omega_\Delta + \Delta)}\right)^2 [\delta(\Delta\omega - \Omega_\Delta + \omega) + \delta(\Delta\omega + \Omega_\Delta - \omega)],\right.$$

$$(4.2.37)$$

where $\Omega_\Delta = \Omega_\Delta(1) \approx (\Delta^2 + 9E_{0x}^2/4)^{1/2}$. The explicit expressions for $I^{(e_x)}(\Delta\omega)$ and $I^{(e_z)}(\Delta\omega)$ depend on the values of the coefficients $d_k^{(c)}$, $d_k^{(s)}$, and Q_k. An analysis of the relations

$$|Q_0| = \max_k |Q_k|, \quad |d_0^{(c)}| = \max_k |d_k^{(c)}|, \quad |d_{\pm 1}^{(s)}| = \max_k |d_k^{(s)}|, \quad |d_0^{(c)}| > |d_{\pm 1}^{(s)}|,$$

which hold under conditions (4.2.16, 22), shows that the most intense feature in the spectrum $I^{(e_x)}(\Delta\omega)$ is the "zero" satellite (at the frequency $\Delta\omega = 0$), while in the $I^{(e_z)}(\Delta\omega)$ spectrum the most intense features are the satellites at the frequencies $\Delta\omega = \pm(\Delta + q\omega) = \pm 3\varepsilon_0/2$. Furthermore, in the $I^{(e_z)}(\Delta\omega)$ spectrum there can be relatively intense satellites at the frequencies $\Delta\omega = 3\varepsilon_0/2 \pm \omega$, $\Delta\omega = -3\varepsilon_0/2 \pm \omega$ and also $\Delta\omega = \pm\omega$.

The fact that the most intense satellites in the $I^{(e_z)}(\Delta\omega)$ spectrum are at the frequencies $\Delta\omega = \pm 3\varepsilon_0/2$ is evidence that at large frequency detuning, $|\Delta| \gg |R_q^{(0)}/2|$, the spectrum $I^{(e_z)}(\Delta\omega)$ is approximately that of a side component in the case of the static Stark effect in a field $\varepsilon_0/2$. The satellites at the frequencies $\Delta\omega = \pm\omega$ share the physical nature of the Baranger–Mozer satellites of the forbidden SL (Sect. 5.1.1). In contrast with the Baranger–Mozer satellites, however, which arise when a dynamic field $E_0 \cos\omega t$, is applied to a two-level system, in the case of the hydrogen atom the satellites at the frequencies $\Delta\omega = \pm\omega$ arise when the atom is subjected to a superposition of a noncollinear static field F and a time-dependent field $E_0 \cos\omega t$. The field F splits the $n = 2$ level into three sublevels, which are separated from each other by $3F$ along the frequency scale; the field $E_0 \cos\omega t$ causes two-quantum transitions to the $n = 1$ level, which give rise to satellites at the frequencies $\Delta\omega = \pm\omega$ on the L_α line. Similar satellites appear on any other hydrogen line $n \to n'$; for $n - n' = 2k - 1(k = 1, 2, \ldots)$ they appear in the $I^{(e_z)}(\Delta\omega)$ spectrum, while for $n - n' = 2k(k = 1, 2, \ldots)$ they appear in the $I^{(e_x)}(\Delta\omega)$ spectrum.

For comparison with the numerical calculations [4.19], we show in Fig. 4.9 the spectrum found for the L_α line from analytic expressions (4.2.35) for several

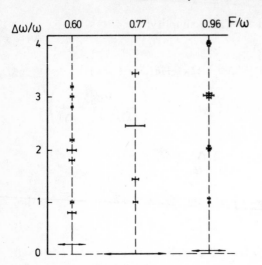

Fig. 4.9. Splitting of the L_α line in nonorthogonal fields F and $E_0 \cos \omega t$ ($E_{0x}/\omega = E_{0z}/\omega = 2$; the z axis runs along F, and the x axis lies in the plane of the vectors F and E_0) for the values of a reduced static field: $F/\omega = 0.60$ (a two-quantum resonance), $F/\omega = 0.77$ (a nonresonant case), and $F/\omega = 0.96$ (a three-quantum resonance). The spectrum $I^{(x)}(\Delta\omega)$ is shown by the line segment with arrowheads, while the $I^{(z)}(\Delta\omega)$ spectrum is shown by the line segments without arrowheads. The length of a segment is proportional to the intensity of the corresponding spectral component

values of the static fields F in the case in which the fields F and $E_0 \cos \omega t$ are not orthogonal. Calculations were carried out for the same absolute value of the dynamic field as in Fig. 2 of [4.19] ($3E_0/\omega = 1.5$), but the vector E_0 had a component along F: $E_{0x} = E_{0z} = 2^{-3/2}\omega$. Under the assumption $P_{\max} = 1/3$, we find from condition (4.2.22) the region in which dynamic resonances exist: $3F/\omega \gtrsim 1.43$. Calculations show that a two-quantum resonance sets in at $3F/\omega \approx 1.8$, and a three-quantum resonance at $3F/\omega \approx 2.9$. For the additional splitting of the SL in the two- and three-quantum resonances we have $\Omega_0(2) = 0.19$ and $\Omega_0(3) = 0.04$, respectively. Comparison with Fig. 2 in [4.19] shows that the angle between F and E_0 has little effect on the positions of the resonances. The reason for this result is that to first order ε_0 depends only on the moduli of the vectors F and E_0.

4.2.4 Dips in Hydrogen Spectral Lines Resulting from the Resonance Effects

To find the resultant profile of the L_α line, we average the spectrum found for this line in Sect. 4.2 over the distribution of quasistatic fields. We consider a very simple model in which the direction of F and the angle between E_0 and F are the same for all of the radiating hydrogen atoms, and the absolute value of the quasistatic field is distributed in accordance with a function $W(F)$. A quasi-one-dimensional static field F can exist in a plasma, for example, when quasi-one-dimensional low-frequency turbulence with a characteristic field $\tilde{F} \gg F_H$ (F_H is the Holtzmark field) develops in the plasma. In this case the resultant profile $S^{(e)}(\Delta\omega)$ is given by

$$S^{(e)}(\Delta\omega) = \int_0^\infty dF\, W(F) I^{(e)}(\Delta\omega). \tag{4.2.38}$$

We are interested primarily in the changes which are caused by the time-dependent field on the resultant profile of a side component which is formed under condition (4.2.23) as a result of the averaging of the spectrum $I^{(e_z)}(\Delta\omega)$ over the distribution $W(F)$ with $e_z \parallel F$. We denote by $F^{(q)}$ that strength of the quasistatic field which corresponds to an exact q-quantum dynamic resonance [$\Delta = 0$ in (4.2.26)].

Let us consider the case of a single-quantum resonance (4.2.36). So that our theory will apply to most of the atoms in a plasma, we require

$$E_0 \ll (F_0, F^{(1)}), \tag{4.2.39}$$

where F_0 is the typical strength of the quasistatic field, and $F^{(1)} = \omega/3$. Furthermore, if the structure due to the single-quantum resonance is to be noticeable on the quasistatic profile of the L_α line, we must require that the function $W(F)$ be substantially nonzero near $F = F^{(1)}$. Since the spectrum $S^{(e_z)}(\Delta\omega)$ is symmetric with respect to $\Delta\omega = 0$, we will consider it only in the blue wing. As follows from (4.2.37, 38), there are three components in the contribution to the intensity of the blue wing in the spectrum $S^{(e_z)}(\Delta\omega)$:

$$S_0^{(e_z)}(\Delta\omega) = \frac{9E_{0x}^2}{8}\delta(\Delta\omega - \omega)\int_0^\infty \frac{dF\, W(F)}{(3F - \omega)^2 + 9E_{0x}^2/4}, \tag{4.2.40}$$

$$S_\pm^{(e_z)}(\Delta\omega) = \frac{81E_{0x}^4}{64}$$

$$\times \int_0^\infty \frac{dF\, W(F)\delta(\Delta\omega - \omega \pm [(3F - \omega)^2 + 9E_{0x}^2/4]^{1/2})}{[(3F - \omega)^2 + 9E_{0x}^2/4]\{[(3F - \omega)^2 + 9E_{0x}^2/4]^{1/2} \pm (3F - \omega)\}^2}, \tag{4.2.41}$$

where $E_{0x} = E_0 \sin\theta$. Performing the integration in (4.2.41), we find

$$S_\pm^{(e_z)}(\Delta\omega) = (3/4)^3 E_{0x}^4 [|\Delta\omega - \omega|R(\Delta\omega)]^{-1}\{W(\omega/3 + R(\Delta\omega)/3)$$

$$\times [\Delta\omega - \omega - R(\Delta\omega)]^{-2} + W(\omega/3 - R(\Delta\omega)/3)$$

$$\times [\Delta\omega - \omega + R(\Delta\omega)]^{-2}\}. \tag{4.2.42}$$

where $R(\Delta\omega) = [(\Delta\omega - \omega)^2 - 9E_{0x}^2/4]^{1/2}$. In (4.2.42), the function $S_-^{(e_z)}(\Delta\omega)$ is defined for $\Delta\omega > \omega + 3E_{0x}/2$ and the function $S_+^{(e_z)}(\Delta\omega)$ for $\Delta\omega < \omega - 3E_{0x}/2$. From (4.2.42) we easily find

$$S_\pm^{(e_z)}(\Delta\omega) \approx \begin{cases} E_{0x}W(\omega/3)/4R(\Delta\omega), & \Delta\omega \approx \omega \pm 3E_{0x}/2, \\ W(\Delta\omega/3)/3, & |\Delta\omega - \omega| \gg 3E_{0x}/2. \end{cases} \tag{4.2.43}$$

The single-quantum resonance (4.2.36) thus leads, under condition (4.2.39), to the following characteristic features in the profile of a lateral component of the L_α line near the frequencies $\Delta\omega = (n_1 - n_2)\omega$, $n_1 - n_2 = \pm 1$, where n_1 and n_2 are parabolic quantum numbers for $n = 2$.

1) There are three peaks near $\Delta\omega = (n_1 - n_2)\omega$ in this spectrum: a central peak at $\Delta\omega = (n_1 - n_2)\omega$ and two side peaks at $\Delta\omega = (n_1 - n_2)\omega \pm 3E_{0x}/2$.

In the model adopted here, the emission in the side peaks does not fall in the frequency interval

$$|\Delta\omega - (n_1 - n_2)\omega| < 3E_{0x}/2. \tag{4.2.44}$$

Far from $\Delta\omega = (n_1 - n_2)\omega$ the spectrum becomes a quasistatic profile of a side component which is described by the function $W(\Delta\omega/3)/3$.

2) If at the resonant value $F = \omega/3$ $W(F)$ is not negligibly small compared to its maximum, then we have a ratio $J_2^{(1)}/J_1^{(1)} < 1$, where $J_1^{(1)}$, $J_2^{(1)}$ are the

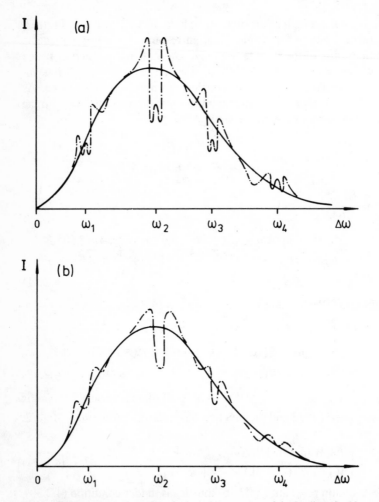

Fig. 4.10. Resultant Stark profile of a lateral component of the L_α line (the blue wing). *Solid line* is the resultant profile in the absence of a dynamic field. The effect of the dynamic field $E_0 \cos\omega t$ is shown by the *dashed line* for four frequencies $\omega = \omega_j$ $(j = 1, 2, 3, 4)$. Near $\Delta\omega = \omega_j$, structures (dips) appear on the resultant profile because of the single-quantum resonance: (**a**) spectral resolution allows "fine structure" of dips to be seen; (**b**) spectral resolution is not sufficient to see the fine structure

intensities of the emission concentrated in the interval (4.2.44) in the absence and presence, respectively, of a one-dimensional time-dependent field ($J_2^{(1)}$ is the intensity of the central peak). If, on the other hand, we have $E_{0x} \to 0$, then we find $J_2^{(1)}/J_1^{(1)} \to \pi/4$ and this result depends on neither the function $W(F)$ nor the frequency ω.

3) The intensity of the emission in the side peaks exceeds that of the emission at the frequencies $\Delta\omega = (n_1 - n_2)\omega \pm 3E_{0x}/2$ in the absence of a time-dependent field.

4) If the central peak cannot be spectrally resolved, a dip appears in the profile of the side component interval (4.2.44) according to conclusions 2 and 3. The relative depth of this dip depends on ω, E_{0x}, and the shape of the function $W(F)$; in principle, this depth can be determined from (4.2.40 and 42). Such a dip may be called a "dip with fine structure".

5) If near $\Delta\omega = (n_1 - n_2)\omega$ we have a derivative $dW(\Delta\omega/3)/d(\Delta\omega) \neq 0$; then yet another dip can exist in this neighborhood. If $dW(\Delta\omega/3)/d(\Delta\omega) > 0$, the dip will be at $\Delta\omega > (n_1 - n_2)\omega + 3E_{0x}/2$, and if $dW(\Delta/3)/d(\Delta\omega) < 0$ it will be at $\Delta\omega < (n_1 - n_2) - 3E_{0x}/2$.

Figure 4.10 shows a representative feature on the quasistatic profile of a side component of the L_α line. The feature stems from a single-quantum resonance with a dynamic field.

An analysis similar to that carried out for the single-resonance shows that for a resonance with an arbitrary number of quanta q ($q = 1, 2, \ldots$) the basic changes in the profile of a side component of the L_α line occur near the frequencies $\Delta\omega = q(n_1 - n_2)\omega$ (resonant splitting appears). The qualitative nature of this splitting does not depend on q. The magnitude of the resonant splitting (the distance between the central and side peaks) is $|R_q^{(0)}|/2$. If the time-dependent field has a small amplitude, $E_0 \ll \omega$, the relation $|R_q^{(0)}| \propto E_0^q$ holds, so that in the case $E_0 \ll \omega$ there is a noticeable splitting only for $q = 1$. If $E_0 \gtrsim \omega$, however, the single-quantum resonance disappears [applicability condition (4.2.22) is violated]. On the other hand, the condition $E_0 \gtrsim \omega$ is a necessary condition for resonant splitting to be noticeable in the case of a q-quantum resonance ($q \neq 1$).

4.2.5 Intra-Stark Spectroscopy. Diagnostic Recommendations

It can be asserted on the basis of these results that if, for some group of hydrogen atoms in a plasma the condition for a q_α-quantum resonance holds for a level n_α, while for some other group of hydrogen atoms the condition for a q_β-quantum resonance holds for a level $n_\beta < n_\alpha$, then a resonance feature will appear on the resultant profile of a side component of the hydrogen SL corresponding to the transition $n_\alpha \to n_\beta$. This feature will occur at the following distances from the center of the line:

$$\omega_\alpha = n_\alpha^{-1} q_\alpha \omega X_{\alpha\beta}, \quad \omega_\beta = n_\beta^{-1} q_\beta \omega X_{\alpha\beta};$$
$$X_{\alpha\beta} \equiv n_\alpha (n_1 - n_2)_\alpha - n_\beta (n_1 - n_2)_\beta. \tag{4.2.45}$$

According to (4.2.18), the number of peaks in the resonant feature near the frequencies ω_α and ω_β does not exceed $2n_\alpha - 1$ and $2n_\beta - 1$, respectively.

The relative depth of the dips apparently increases if $\omega_\alpha = \omega_\beta$. This condition may hold if there is a simultaneous resonance for the upper and lower levels: $n_\alpha q_\beta = n_\beta q_\alpha$. Such "superimposed" dips could conveniently be detected by making use (for example) of the H_β line, which lacks a central component: $n_\alpha = 4$, $n_\beta = 2$, $q_\alpha = 2$, $q_\beta = 1$.

The dips or depressions on Stark profiles of hydrogen SL caused by joint resonant action of static and dynamic EF were observed and analyzed in many experiments carried out on various installations in different countries. Figures 4.11 and 12 display some relevant experimental results obtained in the 1970s.

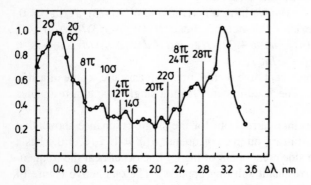

Fig. 4.11. Experimental profile of the H_δ line from [4.20]. The *vertical lines* mark the theoretically expected positions of depressions (dips). The Stark π- and σ-components (to which dips belong) are indicated

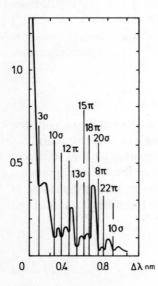

Fig. 4.12. The experimental profile of the H_γ line from [4.21]. The notations are the same as in Fig. 4.11

A relatively new branch of plasma spectroscopy has arisen: spectroscopy within the static Stark profile or intra-Stark spectroscopy (just as intra-Doppler spectroscopy is now a branch of nonlinear optics). At present diagnostics methods based on intra-Stark spectroscopy are widely used Sects. 7.2–4).

The structural features in the profiles of spectral lines which we have studied here can be used to diagnose plasmas with linearly polarized QEFs. By comparing the relative distances of the dips observed in the profile of a spectral line with the set of values $(q_\alpha X_{\alpha\beta}/n_\alpha, q_\beta X_{\alpha\beta}/n_\beta)$, we can find the numbers q_α and q_β and also the frequency ω (if ω is not known at the outset). If we then measure the characteristic halfwidths of the dips, we can use (4.2.18) to determine $R^{(0)}_{q_\alpha}(E_0, F)$ and $R^{(0)}_{q_\beta}(E_0, F)$. Employing these values and making use of the resonance condition (4.2.12), $3n_\nu\varepsilon_0(E_0, f)/4 = q_\nu\omega$ $(\nu = \alpha, \beta)$, we can find that amplitude E_0 of the dynamic field and that value F of the quasistatic field for which resonances occur for the upper and lower levels. If the fine structure of a feature cannot be resolved, it should be possible to observe an ordinary dip with an averaged halfwidth

$$(\Delta\omega^{\mathrm{dip}}_{1/2})_\nu = \langle|n_1 - n_2|_\nu\rangle R^{(0)}_{q_\nu}(E_0, F)/2,$$
$$\langle|n_1 - n_2|_\nu\rangle \approx n_\nu/2 \quad (\nu = \alpha, \beta). \tag{4.2.46}$$

We wish to stress that all the results can be extended without difficulty to a plasma with multimode (but linearly polarized) QEF, $E(t) = \sum_j E_j \cos(\omega t + \varphi_j)$. It is sufficient to average the results derived above for the single-mode case over a Rayleigh distribution of the amplitudes E_0:

$$W_R(E_0, \bar{E}) = (E_0/\bar{E}^2)\exp(-E_0^2/2\bar{E}^2), \quad \bar{E} = \left(\sum_j E_j^2/2\right)^{1/2}.$$

In particular, for a single-quantum resonance the expression for the half-width of a dip (averaged over its fine structure) becomes

$$(\Delta\omega^{\mathrm{dip}}_{1/2})_\nu \approx (3n_\nu\bar{E}/4)\langle|n_1 - n_2|_\nu\rangle, \quad \langle|n_1 - n_2|_\nu\rangle \approx n_\nu/2 \quad (\nu = \alpha, \beta). \tag{4.2.47}$$

Note that the multiquantum dynamic resonance can be used to measure the parameters of quasimonochromatic electric fields in plasmas which are stronger $(E_0/\omega > 1$ or $\bar{E}/\omega > 1)$ than in the case of a single-quantum resonance.

This new area of plasma spectroscopy (intra-Stark spectroscopy) has stirred a lot of theoretical and experimental interest. While experimental results are presented in Chap. 7, let us address here some theoretical comments or follow-ups [4.24–26].

Griem [4.24] contends that in dense plasmas ($N_e \gtrsim 10^{18}$ cm^{-3}) electron impact broadening should make it impossible to observe dips in the H$_\alpha$ profile (in distinction from the dips in the L$_\alpha$ profile). It is true that with an increase of the principal quantum number n, electron impact broadening brings up limits for observations of dips. However, in the phenomenon of dips we deal with electron impact broadening of *quasienergy states* (QS) rather than of usual states. In other words, it is electron impact broadening of a hydrogen atom "dressed" by a QEF $E_0 \cos \omega t$. Estimates based on our theory presented in Sect. 3.3.1 show that at $N_e \gtrsim 10^{18}$ cm^{-3} the electron impact width of the QS involved into the formation of dips in hydrogen line profiles is significantly smaller than the electron impact width of the usual states of hydrogen that was improperly used by Griem. No wonder that structures observed in the H_α profiles can be consistently interpreted as the dips for some experiments [4.27].

Griem [4.24] also contends that "ion-dynamical effects probably reduce the amplitude of profile modulations calculated on the basis of quasistatic ion broadening by another factor $\leqslant 2$". This might be true for early experiments where the dips were observed at densities $N_e = (3 \times 10^{13} - 3 \times 10^{15})$ cm^{-3} [see Chap. 7]. However, it is well known that for hydrogen lines radiated by plasmas of significantly higher densities ($N_e \gtrsim 10^{18}$ cm^{-3}) ion-dynamical effects are much less important. Moreover, it follows from the theory [4.28] that for the dips located in the *wings* of a spectral line (i.e. $\lambda^{\text{dip}} \gg \Delta\lambda_{1/2}$) ion-dynamical effects are further diminished by a factor $\sim \Delta\lambda_{1/2}/\lambda^{\text{dip}} \ll 1$.

Günter and Könies [4.25], referring to our theory of the dips [4.3,27] allege: "No statements could be made about their extent in the detuning. Unfortunately, this is crucial to answering the question of the experimental visibility". However, in the reality analytical formulas for the dip's extent in the detuning (i.e., for the width of the dips) were published in both our paper [4.3] (in its Sect. 5) and in our paper [4.27] (formula (20)) [see also formulas (4.2.46), (4.2.47) above]. It is indeed unfortunate that Günter and Könies failed to read properly even the papers [4.3,27] they refer to in their work [4.25].

A similar case of a reading failure seem to be characteristic also for Fursa and Yudin. Their paper [4.26] contains lots of ridiculous statements like "the consideration in [4.3] was constrained to n = 2", while in the reality in the Sect. 2 of our paper [4.3] a general solution of the problem is obtained for arbitrary n [see also Sect. 4.2.2 above]. Our analysis of their paper [4.26] shows that in fact they have reproduced our published results (known to them) even without extending the validity limit of our results (contrary to their claim).

4.3 Hydrogen-like Spectral Lines in a High-Frequency or Strong QEF with a Quasistatic EF

4.3.1 Calculation of Quasienergy States

According to the general results of Sect. 2.3, quasienergies of a Coulomb radiator in an EF

$$E(t) = E_0 \cos \omega t + F \tag{4.3.1}$$

are given by the eigenvalues of a stationary matrix

$$\langle\langle On\alpha'|V|On\alpha\rangle\rangle = \langle n_1'n_2'm'|x F_x J_0(3n E_0/(2Z\omega)) + z F_z|n_1n_2m\rangle \tag{4.3.2}$$

and periodic parts $\tilde{\varphi}_{n\alpha}$ of correct states of zeroth order (CSZO) are the linear combinations of states

$$|On\alpha\rangle\rangle = |n_1 n_2 m\rangle \exp[-(2Z\omega)^{-1}3in(n_1 - n_2)E_0 \sin \omega t] \qquad (4.3.3)$$

with coefficients determined by the eigenvectors of matrix (4.3.2). Using the ideas presented in [4.4] we can write out analytically the QSs for any n.

Indeed, a matrix coinciding with (4.3.2) is obtained under the consideration of a Stark effect in the following *static* field

$$F_{\text{eff}}^{(n)} = (F_x J_0(3nE_0/2Z\omega), 0, F_z) \qquad (4.3.4)$$

in the basis $|n_1 n_2 m\rangle$ with the quantization axis $Oz \parallel E_0$. But eigenvalues and eigenvectors for a static EF $F_{\text{eff}}^{(n)}$ are known, namely, the states of parabolic quantization $|n'_1 n'_2 m'\rangle$ with the *quantization axis* $Oz' \parallel F_{\text{eff}}^{(n)}$ and with energies (relative to ε_n)

$$\tilde{\varepsilon}_{n'_1 n'_2 m'} = 3n(n'_1 - n'_2)F_{\text{eff}}^{(n)}/2Z. \qquad (4.3.5)$$

Therefore the problem is reduced to the calculation of the matrix $\langle n_1 n_2 m | n'_1 n'_2 m' \rangle$ which transforms parabolic WFs upon rotation of the quantization axis from E_0 to $F_{\text{eff}}^{(n)}$.

We write the corresponding QS for any n:

$$\Psi_{n'_1 n'_2 m'} = \exp(-i\varepsilon_n t - i\tilde{\varepsilon}_{n'_1 n'_2 m'}t) \sum_{\substack{n_1, n_2, m \\ (n_1 + n_2 + |m| + 1 = n)}}$$
$$\times \left[\langle n_1 n_2 m | n'_1 n'_2 m' \rangle | n_1 n_2 m\rangle \exp\left(-\frac{3in(n_1 - n_2)E_0}{2Z\omega} \sin \omega t \right) \right]. \qquad (4.3.6)$$

The validity conditions for the obtained results are the following. We have neglected the mixing of states with different principal quantum numbers, which is correct under the condition

$$\max(n^2 F/Z, n^2 E_0/Z, \omega) \ll |\varepsilon_n - \varepsilon_{n+1}| \sim Z^2/n^3. \qquad (4.3.7)$$

We have also omitted the first order corrections (2.2.14) to QSs which are relatively small when

$$\max(\omega, (Z^{-1}nE_0\omega)^{1/2}) \gg n^2 F/Z \qquad (4.3.8)$$

Note that at $E_0 \to 0$ (4.3.5,6) describe the usual linear Stark effect in a static field F.

4.3.2 Calculations of the L_α, L_β and H_α Line Profiles

The spectra are calculated as a sum of intensities of radiative transitions (1.5) between all QSs of upper and lower levels (assuming all QSs of an upper level to be equally populated). The results for L_α in two polarizations are the following:

$$I^{(e_z)}(\Delta\omega) = \sum_{k=-\infty}^{+\infty} \{\delta(\Delta\omega - 2k\omega)2(\sin^2\varphi_2)J_{2k}^2(3\beta)$$

$$+ [\delta(\Delta\omega - 2k\omega - 3F_{\text{eff}}^{(2)}/Z) + \delta(\Delta\omega - 2k\omega + 3F_{\text{eff}}^{(2)}/Z)]$$

$$\times J_{2k}^2(3\beta)\cos^2\varphi_2 + [\delta(\Delta\omega - (2k-1)\omega - 3F_{\text{eff}}^{(2)}/Z)$$

$$+ \delta(\Delta\omega - (2k-1)\omega + 3F_{\text{eff}}^{(2)}/Z)]J_{2k-1}^2(3\beta)\cos^2\varphi_2,$$

$$I^{(e_x)}(\Delta\omega) = \delta(\Delta\omega)(\cos^2\varphi_2)/2$$

$$+ [\delta(\Delta\omega - 3F_{\text{eff}}^{(2)}/Z) + \delta(\Delta\omega - 3F_{\text{eff}}^{(2)}/Z)]\sin^2\varphi_2;$$

$$\beta \equiv E_0/(\omega Z), \quad \varphi_n \equiv \arccos(F_z/F_{\text{eff}}^{(n)}). \tag{4.3.9}$$

For the L_β line we show only the z polarization spectrum:

$$I^{(e_z)}(\Delta\omega) = \sum_{j=1}^{2} \sum_{k=-\infty}^{+\infty} \{\delta(\Delta\omega - (2k+1)\omega)I_0^{(2k+1)}/2$$

$$+ [\delta(\Delta\omega - 2k\omega - 9jF_{\text{eff}}^{(3)}/2Z) + \delta(\Delta\omega - 2k\omega$$

$$+ 9jF_{\text{eff}}^{(3)}/2Z]I_j^{(2k)} + [\delta(\Delta\omega - (2k+1)\omega - 9jF_{\text{eff}}^{(3)}/2Z)$$

$$+ \delta(\Delta\omega - (2k+1)\omega + 9jF_{\text{eff}}^{(3)}/2Z]I_j^{(2k+1)};$$

$$I_0^{(2k+1)} \equiv (3/2)J_{2k+1}(9\beta)\sin^4\varphi_3, \quad I_1^{(2k+1)} \equiv J_{2k+1}^2(9\beta)\sin^2\varphi_3\cos^2\varphi_3,$$

$$I_1^{(2k)} \equiv J_{2k}^2(9\beta)\sin^2\varphi_3, \quad I_2^{(2k+1)} \equiv (1/4)J_{2k+1}^2(9\beta)(1+\cos^2\varphi_3)^2,$$

$$I_2^{(2k)} \equiv J_{2k}^2(9\beta)\cos^2\varphi_3. \tag{4.3.10}$$

A fragment of the spectrum (4.3.10) at $\beta = 0.1$ and $\varphi_3 = 90°$ is shown in Fig. 4.13. The spectrum consists of the main components near the unperturbed

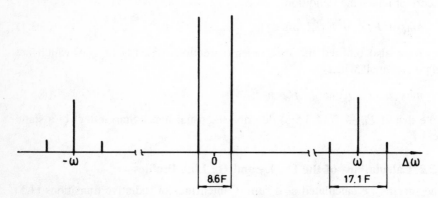

Fig. 4.13. A fragment of the L_β spectrum with z polarization in a field $F + E_0 \cdot \cos\omega t$ at $E_0/Z\omega = 0.1$ and $F \perp E (0z \parallel E_0)$

frequency [corresponding to $k = 0$ in (4.3.10)] and the satellites separated by a multiple of ω [only the first odd satellites, corresponding to $2k + 1 = \pm 1$ in (4.3.10), are shown].

The spectrum of H_α with z polarization consists generally of 15 main components $I_i^{(0)}(i = 0, \pm 1, \ldots, \pm 7)$ and their satellites $I_i^{(k)}$ at frequencies $k\omega$ away $(k = \pm 1, \pm 2, \ldots)$. The components of each satellite are symmetrically located with respect to the central component $I_0^{(k)}$ at the distances

$$\Delta\omega_1 = -\Delta\omega_{-1} = (9F_{\text{eff}}^{(3)}/2 - 3F_{\text{eff}}^{(2)})/Z,$$

$$\Delta\omega_2 = -\Delta\omega_{-2} = 3F_{\text{eff}}^{(2)}/Z, \qquad \Delta\omega_3 = -\Delta\omega_{-3} = 9F_{\text{eff}}^{(3)}/2Z,$$

$$\Delta\omega_4 = -\Delta\omega_{-4} = (9F_{\text{eff}}^{(3)} - 3F_{\text{eff}}^{(2)})/Z,$$

$$\Delta\omega_5 = -\Delta\omega_{-5} = (9F_{\text{eff}}^{(3)}/2 + 3F_{\text{eff}}^{(2)})/Z,$$

$$\Delta\omega_6 = -\Delta\omega_{-6} = 9F_{\text{eff}}^{(3)}/Z, \qquad \Delta\omega_7 = -\Delta\omega_{-7} = (9F_{\text{eff}}^{(3)} + 3F_{\text{eff}}^{(2)})/Z.$$

$$(4.3.11)$$

The intensities of satellites depend on β, φ_2, φ_3 and are generally expressed by rather cumbersome formulas. We write the intensity of even satellites of the central component $(\Delta\omega = 2k\omega, k = 0, \pm 1, \ldots)$:

$$
\begin{aligned}
I_0^{(2k)} = {}& \{[(41/4)J_{2k}(6\beta) + (1/4)J_{2k}(12\beta) + (27/2)J_{2k}(3\beta)]\sin^2\varphi_3 \sin\varphi_2 \\
& + 24J_{2k}(9\beta/2)(\sin\varphi_3)(1 + \cos\varphi_3\cos\varphi_2)\}^2 \\
& + \{[(41/4)J_{2k}(6\beta) + (1/4)J_{2k}(12\beta) + (27/2)J_{2k}(3\beta)] \\
& \times \sin^2\varphi_3 \sin\varphi_2 - 24J_{2k}(9\beta/2)(\sin\varphi_3)(1 - \cos\varphi_3\cos\varphi_2)\}^2 \\
& + \{[(41/2)J_{2k}(6\beta) + (1/2)J_{2k}(12\beta)](\sin^2\varphi_3)\sin\varphi_2 - 27J_{2k}(3\beta) \\
& \times (\sin\varphi_2)\cos^2\varphi_3 + 48J_{2k}(9\beta/2)(\sin\varphi_3)(\cos\varphi_3)\cos\varphi_2\}^2.
\end{aligned}
$$

$$(4.3.12)$$

Note that in contrast to L_β, the H_α spectrum (as well as the L_α spectrum) has no odd satellites of the central component. In Fig. 4.14 a fragment of the H_α spectrum at $\beta = 0.3$ and $\varphi_2 = \varphi_3 = 90°$ is shown.

At $E_0 \to 0$ all the results obtained tend to the usual pattern of the static Stark effect; at $F \to 0$ they are transformed to (3.1.9).

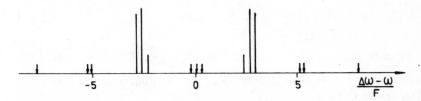

Fig. 4.14. Components of the first satellite of the H_α line in the z polarization in a field $F + E_0 \cos \omega t$ at $E_0/Z\omega = 0.3$ and $F \perp E_0(0z \parallel E_0)$. *Arrows* indicate the components whose intensity (for this satellite) is equal to zero

4.3.3 Further Generalizations for the Action of a Static Magnetic Field

Consider a Coulomb radiator under the simultaneous action of a QEF $E(t) = (0, 0, E_0 \cos \omega t)$ and a static magnetic field $H = (H_x, 0, H_z)$. Using the general results of Sect. 2.3 or the averaging principle [4.16] this dynamic problem may be reduced to a static problem of a Coulomb radiator in an effective magnetic field

$$H_{\text{eff}}^{(n)} = (H_x J_0(3nE_0/(2\omega Z)), 0, H_z) \tag{4.3.13}$$

under the condition

$$(2m\omega)^{-1}n\mu_0|H_x J_m(3nE_0/(2\omega Z))| \ll 1, \quad m = 1, 2, \ldots \tag{4.3.14}$$

(μ_0 is the Bohr magneton). From (4.3.14) the less strict but physically more clear condition

$$\max(\omega, (Z^{-1}n\omega E_0)^{1/2}) \gg n\mu_0|H_x| \tag{4.3.15}$$

can be obtained.
The solution is the following QSs:

$$\psi_{n_1' n_2' m} = \exp(-i\lambda_{nm'}t) \sum_{i_1,i_2=-j}^{+j} \sum_{p=-\infty}^{+\infty} \exp(-ip\omega t)$$

$$\times J_p((2\omega Z)^{-1}3n(n_1 - n_2)E_0)D_{i_1' i_1}^{j}(0, \beta, 0)$$

$$\times D_{i_2' i_2}^{j}(0, -\beta, 0)\varphi_{n_1 n_2 m}(r) \tag{4.3.16}$$

with quasienergies

$$\lambda_{nm'} = \varepsilon_n + \mu_0 m' H_{\text{eff}}^{(n)}. \tag{4.3.17}$$

Here $D_{mm'}^{j}(0, \beta, 0)$ are Wigner functions, $\beta \equiv \arctan[H_z^{-1}H_x J_0(3nE_0/2\omega Z]$, quantum numbers i_1, i_2, i_1', i_2' are connected with parabolic quantum numbers

$$i_1 = 2^{-1}(m + n_2 - n_1), \qquad i_2 = 2^{-1}(m - n_2 + n_1),$$
$$i_1' = 2^{-1}(m' + n_2' - n_1'), \qquad i_2' = 2^{-1}(m' - n_2' + n_1'), \tag{4.3.18}$$

$\varphi_{n_1 n_2 m}(r)$ are parabolic WFs.

Using (4.3.16, 17), we obtain that in the spontaneous emission spectrum the components arise at the frequencies

$$\Delta\omega_{m,m'}^{(k)} = \mu_0[m H_{\text{eff}}^{(n)} - m' H_{\text{eff}}^{(n)}] + k\omega;$$
$$m = 0, \pm 1, \ldots, \pm(n - 1);$$
$$m' = 0, \pm 1, \ldots, \pm(n' - 1); \quad k = 0, \pm 1, \pm 2, \ldots \tag{4.3.19}$$

As an example we show the expressions for the spectrum of the L_α line (with polarizations x, y, z):

$$I_x(\Delta\omega) = (2 \sin^2 \beta)\delta(\Delta\omega) + (\cos^2 \beta)[\delta(\Delta\omega - M) + \delta(\Delta\omega + M)],$$
$$I_y(\Delta\omega) = \delta(\Delta\omega - M) + \delta(\Delta\omega + M),$$

$$I_z(\Delta\omega) = \sum_{p=-\infty}^{+\infty} \left\{ J_{2p}^2\left(\frac{3E_0}{Z\omega}\right) \{(\sin^2\beta)[\delta(\Delta\omega - M + 2p\omega)\right.$$

$$+ \delta(\Delta\omega + M + 2p\omega)] + (2\cos^2\beta)\delta(\Delta\omega + 2p\omega)\}$$

$$\left. + 2J_{2p-1}^2(3E_0/Z\omega)\delta(\Delta\omega + (2p-1)\omega)\right\};$$

$$M \equiv \mu_0 H_{\text{eff}}^{(2)}, \qquad \beta \equiv \arctan(H_z^{-1}H_x J_0(3E_0/Z\omega)). \tag{4.3.20}$$

The analytical treatment may be generalized to the case in which a Coulomb radiator experiences the action of a QEF $E_0\cos\omega t$ and two static fields: a magnetic field $H = H_\parallel + H_\perp (H_\parallel \parallel E_0, H_\perp \perp E_0)$ and an EF $F = F_\parallel + F_\perp$. In this case the dynamic problem may be also reduced to a static one but with two effective static fields:

$$H_{\text{eff}}^{(n)} = H_\parallel + H_\perp J_0(3nE_0/2\omega Z), \qquad F_{\text{eff}}^{(n)} = F_\parallel + F_\perp J_0(3nE_0/2\omega Z). \tag{4.3.21}$$

In particular, the splitting of the level with principal quantum number n can be written

$$\lambda_{nn'n''} = |\mu_0 H_{\text{eff}}^{(n)} - (2Z)^{-1}3nF_{\text{eff}}^{(n)}|n' + |\mu_0 H_{\text{eff}}^{(n)} + (2Z)^{-1}3nF_{\text{eff}}^{(n)}|n'',$$

$$n', n'' = -j, -j+1, \ldots, j, \quad j = 2^{-1}(n-1). \tag{4.3.22}$$

The results (4.3.21, 22) correspond to the case of a high frequency or a strong QEF $E_0\cos\omega t$

$$\max(\omega, (Z^{-1}n\omega E_0)^{1/2}) \gg \max(n\mu_0 H_\perp, nF_\perp/Z). \tag{4.3.23}$$

The opposite case was analyzed in [4.22], where it was shown that resonance features (depressions or dips) analogous to those discussed in Sect. 4.2 may appear on the SL profiles emitted by an ensemble of hydrogen atoms from a plasma. The components of the emission spectrum of the atoms that enter into a resonance with the upper level n_a are located in the vicinities of the following frequencies

$$\Delta\omega_{\alpha\beta}^{(a)} = n_a^{-1}\omega\{n_a(n' + n'')_\alpha - [(\mu_0 H/\omega)^2(n_a^2 - n_b^2) + n_b^2]^{1/2}(n' + n'')_\beta\};$$

$$n'_\alpha, n''_\alpha = -j_a, -j_a+1, \ldots, j_a; \qquad n'_\beta, n''_\beta = -j_b, -j_b+1, \ldots, j_b;$$

$$j_l = (n_l - 1)/2, \quad l = a, b. \tag{4.3.24}$$

Here it was taken into account that in the field $H + F$ the positions of the spectral components are determined by the frequencies

$$(n' + n'')_\alpha \kappa^{(a)} - (n' + n'')_\beta \kappa^{(b)},$$

$$(\kappa^{(l)} = [(\mu_0 H)^2 + (3n_l F/2)^2]^{1/2}, \quad l = a, b), \tag{4.3.25}$$

with the resonance condition $\kappa^{(a)} = \omega$.

Similarly, in the case of resonance for the lower level n_b, the components are located in the vicinities of the frequencies

$$\Delta\omega_{\alpha\beta}^{(b)} = n_b^{-1}\omega\{[n_a^2 - (\mu_0 H/\omega)^2(n_a^2 - n_b^2)]^{1/2}$$
$$\times (n' + n'')_\alpha - n_b(n' + n'')_\beta\}, \tag{4.3.26}$$

The formulas (4.3.24–26) are valid under the additional assumption

$$3n\mu_0|HF|/2 \ll \max(\mu_0^2 H^2, 9n^2 F^2/4); \quad n = n_a, n_b. \tag{4.3.27}$$

For resonance features to appear on the resultant quasistatic profiles of the SL it is necessary that the value of frequency $\Delta\omega_{\alpha\beta}^{(l)}$ be one and the same for most emitting hydrogen atoms for which the resonance condition $\kappa^{(l)} = \omega$ ($l = a, b$) is met. This is possible, for example, in the case where an ensemble of radiating hydrogen atoms is in a uniform magnetic field H = const or where the magnetic field is not uniform, but on the average $\mu_0 H \gg 3n_l F/2, l = a, b$ (in this case $\mu_0 H \approx \omega$). In addition, in any case the resonance reliefs will exist at frequencies

$$\Delta\omega_{\alpha\beta}^{(a)} = \omega(n' + n'')_\alpha \text{ [i.e. at } (n' + n'')_\beta = 0 \text{ in (4.3.24)] and,}$$

$$\Delta\omega_{\alpha\beta}^{(b)} = -\omega(n' + n'')_\beta \text{ [i.e. at } (n' + n'')_\alpha = 0 \text{ in (4.3.26)].}$$

We note, in conclusion, some useful general scaling laws for these problems presented in [4.23]. The main result obtained in [4.23] is that under the combined action of a time-periodic EF $E(t)$ and a time-periodic magnetic field $H(t)$ the splitting of an arbitrary hydrogen multiplet with principal quantum number n may be inferred from the corresponding information for $n' = 2$ by scaling E by $n/2$ (so that $nE = n'E'$) and leaving H unchanged.

The n^2 distinct quasienergies $\lambda_{n\mu^+\mu^-}$ should generally be

$$\lambda_{n\mu^+\mu^-} = \varepsilon_n + \mu^+ k_n^+ + \mu^- k_n^-,$$

$$\mu^+, \mu^- = -(n-1)/2, -(n-3)/2, \ldots, (n-1)/2. \tag{4.3.28}$$

In some particular cases the splitting of an n-multiplet may be simplified to $2n - 1$ equally spaced quasienergies

$$\lambda_{n\mu} = \varepsilon_n + \mu k_n; \quad \mu = -(n-1), \ldots, 0, \ldots, n-1. \tag{4.3.29}$$

This happens in the following cases: (1) pure magnetic field ($E = 0$); (2) pure EF confined to a plane; (3) planar EF $E(t)$ and magnetic field $H(t) \perp E(t)$ for all times.

It should be mentioned that the scaling law from [4.24] may be further generalized to the case of hydrogen-like ions with nuclear charge Z: the splitting of an arbitrary n-multiplet can be obtained from the corresponding splitting of hydrogen for $n' = 2$ by scaling E by $n/2Z$ (so that $nE/Z = n'E'$) and leaving H unchanged.

5 Action of a One-Dimensional QEF on Non-Coulomb Emitters

This chapter discusses the satellites of dipole-forbidden SL of emitters (radiators) such as helium and alkali atoms (as well as helium-like and alkali-like ions). In the energy spectra of these radiators quasi-isolated subsystems consisting of two or three levels (inside a multiplet) that are most strongly mixed by the EF can be singled out, while other levels may be neglected. Special emphasis is placed on methods of theoretical description beyond the perturbation theory. In Sects. 5.3.2 and 5.5 the analogous effect in spectra of hydrogen-like ions and diatomic polar molecules is investigated. In contrast to all other sections, in Sect. 5.4 we analyze not the direct action of QEF on a radiator but the influence caused by the changes induced by the QEF in the plasma electron distribution functions. In Sect. 5.6 the case of joint action of a resonant laser field and low-frequency QEF is treated, with relevant applications to laser spectroscopy pointed out. The main results of Chap. 5 are contained in [5.1–11].

5.1 Satellites of Dipole-Forbidden Spectral Lines Caused by a Nonresonant Action of QEFs (Three-Level Scheme)

5.1.1 Dirac Perturbation Theory. Baranger-Mozer Method for Measurements of QEF Parameters

The first theoretical work devoted to the diagnostics of QEF in plasmas by SL of non-Coulomb radiators was that of *Baranger and Mozer* [5.1]. They singled out, in the energy spectrum of a non-Coulomb emitter (a helium atom) the system of three levels 0, 1, 2 with the following properties (Fig. 5.1): the neighboring levels 1, 2 are connected by a dipole matrix element; the radiative transition from the level 2 to some distant level 0 is dipole-allowed, but from the level 1 to the level 0 it is dipole-forbidden. It was known that under the action of a static EF F in a spontaneous emission spectrum of this system, besides an allowed SL at the frequency $\omega_{20} = \omega_2 - \omega_0$ with an intensity I_a, a forbidden SL appears at the frequency $\omega_{10} = \omega_1 - \omega_0$ with an intensity $I_F \propto F^2$ (at small F; here ω_0, ω_1, ω_2 are the energies of the levels 0, 1, 2 in atomic units). In [5.1] it was shown that under the action of QEF, at a frequency ω in the spontaneous emission spectrum, instead of a forbidden SL two of its satellites must appear at the frequencies $\omega_{sat} = \omega_{10} \pm \omega$.

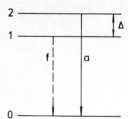

Fig. 5.1. The scheme of dipole-allowed (a) and dipole-forbidden (f) transitions in a three-level system which is selected from the energy spectrum of non-Coulomb radiators

The calculations in [5.1] were carried out for the case of an isotropic multimode QEF. Within the usual nonstationary perturbation theory (Dirac PT) the following expression for satellite intensities S_+, S_- was obtained ($\hbar = m_e = e = 1$):

$$S_\pm/I_a = [6(\Delta \pm \omega)^2(2l_2 + 1)]^{-1}\langle E^2 \rangle \max(l_1, l_2) \left[\int_0^\infty R_{l_1}(r) R_{l_2}(r) r^3 \, dr \right]^2 ,$$

(5.1.1)

Here $\langle E^2 \rangle$ is the root-mean-square amplitude of the QEF, Δ is the separation between the levels 1 and 2, l_1 and l_2 are their orbital quantum numbers, $R_{l_1}(r)$ and $R_{l_2}(r)$ are the radial parts of the spherical WF.

Later, in [5.2], the intensities of analogous satellites caused by the presence (in a plasma) of a linearly polarized singlemode QEF (in particular, a laser or maser field) of the form $E(t) = E_0 \cos \omega t$ were calculated taking into account polarization characteristics. The result obtained there within the Dirac PT may be presented by the following formula (for observations transverse to the axis $Oz \| E_0$):

$$\frac{S_\pm}{I_a} = \frac{E_0^2}{4(\Delta \pm \omega)^2} \frac{\sum\limits_{m_0, m_1, m_2} |z_{12}|^2(|y_{20}|^2 + |z_{20}|^2)}{\sum\limits_{m_0, m_2} (|y_{20}|^2 + |z_{20}|^2)}.$$

(5.1.2)

To each of the levels 0, 1, 2 several WFs differing by magnetic quantum numbers m_0, m_1, m_2 may belong; that is why the corresponding summations appear in (5.1.2). The validity of the results of [5.1.2] is restricted by two conditions: 1) the absence of resonances of the type $q\omega \approx \Delta$; 2) the relative weakness of the QEF $-\max(\Delta, \omega) \gg |r_{12}|(\langle E^2 \rangle)^{1/2}$, $|r_{12}|E_0$.

Baranger and Mozer proposed to measure a QEF amplitude E_0 [or $(\langle E^2 \rangle)^{1/2}$] by comparing the experimentally obtained ratio $S_\pm/I_a$ with the theoretical dependence $S_\pm/I_a$ on E_0 [or on $(\langle E^2 \rangle)^{1/2}$]. Since *Kunze and Griem* [5.12] first did this experiment, the same idea for QEF measurements has been used many times in plasma studies. But the experiments have gradually shifted into the regime of relatively strong fields for which Dirac PT is inapplicable ($2E_0z_{12}/\Delta \equiv \alpha \gtrsim 1$).

Most of these experiments are in the low-frequency regime $\omega \ll \Delta$. Then the problem may be solved analytically using the adiabatic PT (Sect. 2.1), which is valid for $\alpha\omega/\Delta \ll 1$, i.e. for the case $\alpha \gtrsim 1$ also. The corresponding analytical solution [5.3, 4] (discussed below) has allowed to extend the Baranger–Mozer idea to the strong fields region and to develop a new diagnostics method.

5.1.2 Adiabatic Theory of Satellites and Quasilocal Method for Measurements of QEF Parameters

For the three-level system 0, 1, 2 shown in Fig. 5.1, in a linearly polarized QEF $E(t) = E_0 \cos \omega t$ the Schrödinger equation can be written

$$i\,\partial\psi/\partial t = (H_a + zE_0 \cos \omega t)\psi, \qquad (5.1.3)$$

where H_a is the Hamiltonian of an isolated atom. To obtain results valid in stronger fields, we use the adiabatic PT for the subsystem of the two levels 1 and 2. It is known [5.13] that the instantaneous eigenvalues of the operator $H(t) = H_a + zE_0 \cos \omega t$ are

$$\omega_{1,2}(t) = [\omega_1^{(0)} + \omega_2^{(0)} \mp \Delta(1 + \alpha^2 \cos^2 \omega t)^{1/2}]/2. \qquad (5.1.4)$$

These eigenvalues correspond to the wave eigenfunctions (adiabatic basis)

$$\chi_1(t) = \psi_1 \cos(\beta/2) - \psi_2 \sin(\beta/2),$$
$$\chi_2(t) = \psi_1 \sin(\beta/2) + \psi_2 \cos(\beta/2),$$
$$\beta \equiv \arctan(\alpha \cos \omega t). \qquad (5.1.5)$$

We seek the solution of (5.1.3) in the form

$$\psi(t) = \sum_{j=1}^{2} C_j(t)\chi_j(t) \exp\left[-i \int_0^t dt'\, \omega_j(t')\right]. \qquad (5.1.6)$$

Substitution of (5.1.6) into (5.1.3) yields

$$\dot{C}_1 = -C_2(\dot{\beta}/2) \exp\left[-i \int_0^t dt'\, \omega_{21}(t')\right]$$

$$\dot{C}_2 = C_1(\dot{\beta}/2) \exp\left[i \int_0^t dt'\, \omega_{21}(t')\right]; \qquad (5.1.7)$$

$$\dot{\beta} = -\alpha\omega(\sin \omega t)/(1 + \alpha^2 \cos^2 \omega t). \qquad (5.1.8)$$

We solve the system (5.1.7) by PT. To calculate the satellite intensities, the initial conditions $C_1(0) = 1$ and $C_2(0) = 0$ suffice. We then get from (5.1.7)

$$C_2(t) \approx \frac{1}{2} \int_0^t dt'\, \dot{\beta}(t') \exp\left(i \int_0^{t'} d\tau\, \omega_{21}(t)\right). \qquad (5.1.9)$$

To determine the integrals in (5.1.9) we expand $\dot{\beta}(t)$ and $\omega_{21}(t)$ in a Fourier series:

$$\dot{\beta}(t) = \mathrm{i}(k\omega/4) \sum_{p=-\infty}^{+\infty} (a_{2p} - a_{2p+2}) \exp[\mathrm{i}(2p+1)\omega t],$$

$$\omega_{21}(t) = \bar{\Delta} + \sum_{q=1}^{\infty} \varepsilon_{2q} \cos 2q\omega,$$

$$\bar{\Delta} \equiv (2/\pi)\Delta(1+\alpha^2)^{1/2}\mathbb{E}(k), \tag{5.1.10}$$

where $k \equiv \alpha(1+\alpha)^{-1/2}$, $a_{2p} = 2(-1)^p k^{-2|p|}[(1-k^2)^{1/2} - 1]^{2|p|}$, $\varepsilon_2 = (4/3\pi)\Delta(1+\alpha^2)^{1/2}[\mathbb{E}(k) - 2(1-k^2)\mathbb{D}(k)]$, $\varepsilon_4 = (4(15\pi k^4)^{-1}\Delta \times (1+\alpha^2)^{1/2}[(8k^4-24k^2+16)\mathbb{K}(k)+(-k^4+16k^2-16)\mathbb{E}(k)]$; $\mathbb{E}(k)$, $\mathbb{D}(k)$, $\mathbb{K}(k)$ – are complete elliptic integrals.[1] We consider hereafter only the case $\alpha \lesssim 1$, when it suffices to retain in $\dot{\beta}(t)$ only the terms with $a_0 = 2$ and a_2. In the expansion of the exponential in (5.1.9) we confine ourselves to the expression

$$\exp\left[\mathrm{i}\int_0^t \mathrm{d}\tau\omega_{21}(\tau)\right] \approx \left[J_0\left(\frac{\varepsilon_2}{2\omega}\right) + J_1\left(\frac{\varepsilon_2}{2\omega}\right)\exp(2\mathrm{i}\omega t)\right.$$

$$\left. -J_1\left(\frac{\varepsilon_2}{2\omega}\right)\exp \times (-\mathrm{i}2\omega t)\right]\exp(\mathrm{i}t\bar{\Delta}). \tag{5.1.11}$$

Taking (5.1.10, 11) into account we get from (5.1.9)

$$C_2(t) \approx (k\omega/8)\{[(2-a_2)J_0 - 2J_1]g(\bar{\Delta}+\omega) - [(2-a_2)J_0 + 2J_1]$$
$$\times g(\bar{\Delta}-\omega)+(2J_1+a_2J_0)g(\bar{\Delta}+3\omega)+(2J_1-a_2J_0)g(\bar{\Delta}-3\omega)\},$$
$$g(u) \equiv [\exp(\mathrm{i}ut) - 1]/u, \tag{5.1.12}$$

where the argument $\varepsilon_2/2\omega$ of the Bessel functions J_0 and J_1 is left out for brevity.

In the first nonvanishing order of the adiabatic PT the WF of the two-sublevel system is thus

$$\psi(t) \approx \chi_1(t)\exp\left[-\mathrm{i}\int_0^t \mathrm{d}t'\,\omega_1(t')\right] + C_2(t)\chi_2(t)\exp\left[-\mathrm{i}\int_0^t \mathrm{d}t'\,\omega_2(t')\right]. \tag{5.1.13}$$

The spontaneous emission spectrum for the transition into state 0 is calculated from the equation

---

[1] Fourier coefficients ε_{2q} may be written in the general form:

$$\varepsilon_{2q} = \Delta(1+\alpha^2)^{1/2}2(-1)^{q+1}\sum_{r=q}^{\infty}(k/2)^{2r}(2r-3)!!(2r-1)!![(r-q)!(r+q)!]^{-1}, \quad q \geqslant 2.$$

$$I^{(e)}(\Delta\omega) = \lim_{T\to\infty} (2\pi T)^{-1} \left| \int_0^T dt \, \langle \psi(t)|\mathbf{re}|\psi_0\rangle \exp(-it\,\Delta\omega) \right|^2, \quad (5.1.14)$$

where $\mathbf{e}$ is the unit vector of photon polarization.

To calculate the spectrum (5.1.14) we expand in a Fourier series the functions $\sin(\beta/2)$ and $\cos(\beta/2)$ contained in $\chi_{1,2}(t)$:

$$\sin(\beta/2) = \sum_{p=0}^{\infty} B_{2p+1} \cos(2p+1)\omega t,$$

$$\cos(\beta/2) = \sum_{p=0}^{\infty} A_{2p} \cos 2p\omega t. \quad (5.1.15)$$

The dependence of the coefficients A_0, A_2, and B_1 on k^2 are shown in Fig. 5.2. We note that at $k^2 \ll 1$ we have $A_0 \approx 1 - k^2/16$, $B_1 \approx k/2 + 7k^3/64$, $A_2 \approx -k^2/16$. We substitute the WFs from (5.1.13, 14) expressing the exponentials in (5.1.14) in a form similar to (5.1.11) and taking (5.1.4, 12, 15) into account. We then ultimately obtain for the spectra $S_-^{(e)}$ and $S_+^{(e)}$ of the near and far satellites

$$S_{\mp}^{(e)}(\Delta\omega) = \sum_{m,m',m''} \sigma_{\mp} |r_{20}\mathbf{e}|^2 \delta(\Delta\omega - (\omega_{10}^{(0)} + \omega_{20}^{(0)} - \bar{\Delta})/2 \mp \omega), \quad (5.1.16)$$

where $\bar{\Delta}(m,m'')$ is defined by (5.1.10), $r_{20}(m'',m') \equiv \langle \psi_2|\mathbf{r}|\psi_0\rangle$, and for $\sigma_{\mp}(m,m'')$ we have

$$\sigma_- \approx \{(J_0 - J_1)B_1/2 + [(2-a_2)J_0(\varepsilon_2/(2\omega)) + 2J_1(\varepsilon_2/(2\omega))]$$
$$\times J_0 k A_0 \omega/[8(\bar{\Delta} - \omega)]$$
$$- (2A_0J_1 + A_2J_0)J_0(\varepsilon_2/(2\omega))k\omega/[8(\bar{\Delta} + \omega)]\}^2,$$

$$\sigma_+ \approx \{(J_0 + J_1)B_1/2 - [(2-a_2)J_0(\varepsilon_2/(2\omega)) - 2J_1(\varepsilon_2/(2\omega))]$$
$$\times J_0 k A_0 \omega/[8(\bar{\Delta} + \omega)]$$
$$- (2A_0J_1 - A_2J_0)J_0(\varepsilon_2/(2\omega))k\omega/[8(\bar{\Delta} - \omega)]\}^2 \quad (5.1.17)$$

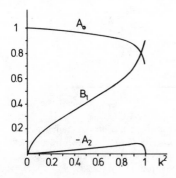

Fig. 5.2. The coefficients A_0, A_2, B_1 of the Fourier expansions (5.1.15) of the adiabatic WFs (5.1.5) vs. the parameter $k^2 \equiv \alpha^2/(1+\alpha^2) = 4z_{12}^2 E_0^2/(\Delta^2 + 4z_{12}^2 E_0^2)$

[the argument of the Bessel functions J_0 and J_1, whenever omitted from (5.1.17) for brevity, is $\varepsilon_2/4\omega$].

The spectrum of the allowed line is

$$I_a^{(e)}(\Delta\omega) \approx \sum_{m,m',m''} (1 - \sigma_- - \sigma_+)|r_{20}e|^2 \delta(\Delta\omega - (\omega_{10}^{(0)} + \omega_{20}^{(0)} + \bar{\Delta})/2).$$

(5.1.18)

We note that in (5.1.16, 18) account is taken of the shift in the satellite position

$$\delta(m, m'') = (\bar{\Delta} - \Delta)/2 = [(2/\pi)(1 + \alpha^2)^{1/2}\mathbb{E}(k) - 1]\Delta/2 \qquad (5.1.19)$$

and of the equal but opposite shift of the allowed line.

The condition for the validity of the approach developed above is the smallness of the coefficient $C_2(t)$ of the WF $\psi(t)$ in (5.1.13): $|C_2(t)| \ll 1$. In the case $\alpha \lesssim 1$ it suffices for this purpose, as seen from (5.1.12), to stipulate that $\omega/\Delta \ll 1$.

Expressions (5.1.16–18) can be used to obtain the intensities of linearly polarized QEFs in a plasma by measuring the ratio $S_\mp/I_a$. The region of their validity, in contrast to [5.1,2], is no longer limited to weak fields with $\alpha \ll 1$. By way of example, Fig. 5.3 shows the analytically calculated plots of S_+/I_a and S_-/I_a vs $E_{rms} = E_0/\sqrt{2}$ for the transitions $(4^1D, 4^1F) \rightarrow 2^1P$ in a helium atom, at two frequencies, $\omega/2\pi c = 0.5$ cm^{-1} and 2 cm^{-1} (the results are summed over the two polarizations). For comparison, the analogous dependences taken from [5.14], where they were obtained numerically, are also shown. (Note that

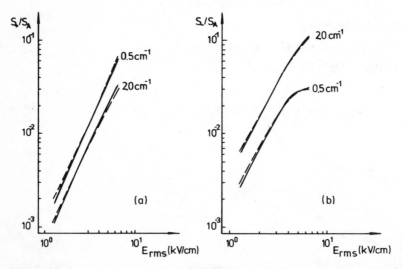

Fig. 5.3. Dependence of the ratio of intensities of the far (**a**) or near (**b**) satellites to the intensity of the allowed line He I 4922 Å on the root-mean-square field $E_{rms} = E_0/2^{1/2}$ for two frequencies, $\omega/2\pi c = 0.5$ cm^{-1} and $\omega/2\pi c = 2$ cm^{-1}. *Solid lines* are analytical results using the approach described here, *dashed lines* are results of computer calculations from [5.13]

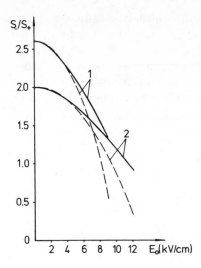

Fig. 5.4. Dependence of the intensity ratio S_-/S_+ of the near and far satellites on the amplitude E_0 of a QEF of frequency $\omega/2\pi c = 1.28$ cm^{-1} for the lines He I 4922 Å and He I 4471 Å. *Solid lines* – results of the adiabatic PT, *dashed lines* – results of Dirac PT (which is valid only at small E_0)

here we have used the more accurate value $\Delta/2\pi c = 5.43$ cm^{-1} from [5.15] instead of 5.63 cm^{-1} from [5.14]). Of course, for another frequency ω used in a different experiment it is also possible to carry out a relatively simple calculation with the aid of the analytic expressions (5.1.16–18).

In a number of cases the volume v_E from which the satellites are emitted can be considerably smaller than the volume V from which the allowed SL is emitted. To determine the QEF intensity it is then possible to measure the satellite intensity ratio S_-/S_+. Figure 5.4 shows plots of S_-/S_+ vs E_0 for the lines He I 4922 A and He I 4471 A, calculated from (5.1.16,17) for the frequency $\omega/2\pi c = 1.28$ cm^{-1}.

We point out that according to (5.1.2), obtained by the Dirac PT with allowance for the terms $\sim E_0^2$, the ratio S_-/S_+ is constant. To obtain the dependence of S_-/S_+ on E_0 in the Dirac PT the calculation must take into account terms at least of the order of E_0^4. Such a calculation leads to the following satellite intensities:

$$\sigma_\pm \approx \frac{1}{(\Delta \pm \omega)^2} \left[\left(\frac{E_0 z_{12}}{2} \right)^2 \pm \left(\frac{E_0 z_{12}}{2} \right)^4 \frac{\Delta^3 \mp 7\Delta^2\omega + 3\Delta\omega^2 \mp \omega^3}{\omega(\Delta^2 - \omega^2)^2} \right].$$
$$(5.1.20)$$

For illustration, the corresponding ratio $(\sum \sigma_-)/(\sum \sigma_+)$ is shown by the dashed lines in Fig. 5.4 (for the same frequency $\omega/2\pi c = 1.28$ cm^{-1}). It can be seen that even when the terms $\sim E_0^4$ are taken into account, a PT calculation is valid only for relatively weak fields.

5.1.3 Polarization of Satellites

The satellite polarization effect can be used to measure the degree of anisotropy of the distribution of QEF directions or the angle between the vector $\boldsymbol{E}(t) =$

$E_0 \cos \omega t$ and the polarizer transmission axis (which is usually determined by the symmetry axis of the experimental setup or by the directions of the polarization and of the wave vector of the external field). We assume that the observation is along the y axis, the polarizer transmission axis is oriented in one case along the z axis and in the other along x, and the vector E is located in the xz plane and makes with the z axis an angle γ which is to be determined. From (5.1.16) for the satellite intensities $S_{\mp}^{(z)}$ (polarization along z) and $S_{\mp}^{(x)}$ (polarization along x) we then obtain

$$S_{\mp}^{(z)} = \sum_{m,m',m''} \sigma_{\mp}(|x_{20}|^2 \sin^2 \gamma + |z_{20}|^2 \cos^2 \gamma),$$

$$S_{\mp}^{(x)} = \sum_{m,m',m''} \sigma_{\mp}(|x_{20}|^2 \cos^2 \gamma + |z_{20}|^2 \sin^2 \gamma). \tag{5.1.21}$$

In practice it is convenient to measure the ratio of the intensities of the same satellite at two orientations of the polarizer:

$$S_{\mp}^{(z)}/S_{\mp}^{(x)} = [1 + f_{\mp}(E_0)\cot^2\gamma]/[\cot^2\gamma + f_{\mp}(E_0)],$$

$$f_{\mp}(E_0) \equiv \left(\sum_{m,m',m''} \sigma_{\mp}|z_{20}|^2 \right) \bigg/ \left(\sum_{m,m',m''} \sigma_{\mp}|x_{20}|^2 \right). \tag{5.1.22}$$

We consider as an example the transitions $(4^1F, 4^1D) \rightarrow 2^1P$. For weak fields we have $\sigma_{\mp} \approx [E_0 z_{12}/2(\Delta \mp \omega)]^2$, and (5.1.22) yields $f_-(E_0) = f_+(E_0) = 4/3$, which agrees with the result of [5.2]. In other words, in weak fields the results of polarization measurements do not depend on the field amplitude. In strong fields the situation is different: the ratio $S_{\mp}^{(z)}/S_{\mp}^{(x)}$ depends not only on the angle γ but also on the field amplitude E_0. This is essentially a new factor not appearing in [5.2].

The functions $f_+(E_0)$ and $f_-(E_0)$ calculated from (5.1.22,17) are shown in Fig. 5.5. If the amplitude E_0 is known and $f_+(E_0)$ or $f_-(E_0)$ is found, we can determine from the experimentally measured ratio $S_{\mp}^{(z)}/S_{\mp}^{(x)}$ the angle γ with the aid of (5.1.22). We note, however, that in strong fields the values of f_-

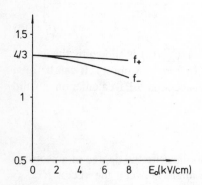

Fig. 5.5. Functions $f_+(E_0)$ and $f_-(E_0)$ from (5.1.22) that determine the intensity ratio of the same satellite at two mutually perpendicular orientations of a polarizer

and f_+ approach unity (the degree of polarization of the satellites decreases), so that the angle γ is less accurately determined from (5.1.22).

5.1.4 Strong Asymmetry of Satellites Distribution in Very Intense QEFs

The problem of measuring very strong QEFs in plasmas is very important. Powerful microwave generators have been developed, in particular, the free-electron laser achieving a peak output of 1 GW at a wavelength of 0.8 cm [5.16]. In experiments on the interaction of a relativistic electron beam with plasma, Langmuir oscillations ($\omega_{pe}/2\pi \approx 64$ GHz) have been detected with EF amplitudes of ~ 100 kV/cm [5.17].

Some of the most useful transitions in spectroscopic diagnostics of very strong QEFs (~ 50–300 kV/cm) in the microwave range are the transitions $(3^1P, 3^1D) \rightarrow 2^1S$, $(3^1P, 3^1D) \rightarrow 2^1P$ of helium. Two essential features emerge in these fields: 1) there is a multisatellite structure in the vicinity of both allowed and forbidden SL even in the situation where matrix elements of the operator of a dipole interaction between the states of levels 3^1P, 3^1D do not exceed the energy splitting between these levels; 2) there is considerable asymmetry of the multisatellite structure in the vicinity of the forbidden SL.

Let us consider in the energy spectrum of an atom a system of three levels 0, 1, 2 where transitions $1 \rightarrow 0$ and $2 \rightarrow 0$ correspond to dipole-forbidden and dipole-allowed SLs, respectively, the close levels 1 and 2 being connected by a dipole transition (Fig. 5.1). In the absence of QEFs, the levels 0, 1, 2 are characterized by the WFs $\varphi_0 = \varphi_{n'l'm'}$, $\varphi_1 = \varphi_{nlm}$, $\varphi_2 = \varphi_{n''l''m''}$ and energies $\omega_0^{(0)}$, $\omega_1^{(0)}$, $\omega_2^{(0)}$. The solution of the Schrödinger equation for the levels 1, 2 in a QEF $E(t) = E_0 \cos \omega t$ (quantization axis Oz is selected along E_0) has the form of adiabatic WFs

$$\psi_s(t) = \chi_s(t) \exp\left[-i \int\limits_0^t dt' \omega_s(t')\right] \quad (s = 1, 2), \tag{5.1.23}$$

where $\omega_s(t)$, $\chi_s(t)$ are given in (5.1.4,5). Expression (5.1.23) holds with the condition $|\omega/\omega_{21}^{(0)}| \ll \min(1, 2/\alpha)$. The spectrum in the vicinity of the allowed SL $I_{2\rightarrow0}^{(e)}(\Delta\omega)$ is due to the transition $\psi_2 \rightarrow \psi_0$ [$\psi_0 = \varphi_0 \exp(-i\omega_0^{(0)}t)$], while the spectrum in the vicinity of the forbidden SL $I_{1\rightarrow0}^{(e)}(\Delta\omega)$ arises from the transition $\psi_1 \rightarrow \psi_0$ (e is the unit vector of polarization of the emitted photons). In calculating these spectra the expansions (5.1.10,15) were used. As a result, under the additional condition $\alpha(1 + \alpha^2)^{-1/2} \ll \min(2^{-1/2}, 4(\omega/\omega_{21}^{(0)})^{1/4})$ the expressions for the spectra could be represented as

$$I_{2\rightarrow0}^{(e)}(\Delta\omega) = \sum_{m',m''} I_{m'',m'}^{(e)}(\Delta\omega), \quad I_{1\rightarrow0}^{(e)}(\Delta\omega) = \sum_{m',m''} \tilde{I}_{m'',m'}^{(e)}(\Delta\omega),$$

$$I_{m'',m'}^{(e)}(\Delta\omega) = A_0^2(m'')|(re)_{20}|^2 \sum_{s=-\infty}^{+\infty} G_{s,2}^2(m'')$$

$$\times \delta(\Delta\omega - \bar{\bar{\omega}}_2(m'') + \bar{\omega}_0(m') - 2s\omega),$$

$$\tilde{I}_{m'',m'}^{(e)}(\Delta\omega) = 4^{-1} B_1(m'') |(re)_{20}|^2 \sum_{s=-\infty}^{+\infty} [G_{s,1}(m'') + G_{s+1,1}(m'')]^2$$

$$\times \delta(\Delta\omega - \bar{\bar{\omega}}_1(m'') + \bar{\omega}_0(m') + (2s+1)\omega), \qquad (5.1.24)$$

where

$$(r,e)_{20} = \langle \varphi_2(m'') | re | \varphi_0(m') \rangle, \quad G_{s,p} = \sum_{r=-\infty}^{+\infty} J_{s-2r}(\varepsilon_2^{(p)}/4\omega) J_r(\varepsilon_4/8\omega),$$

$$\bar{\bar{\omega}}_p = \bar{\omega}_p + \Delta_p, \quad \varepsilon_2^{(p)} = \varepsilon_2 + \Delta_p, \quad (p = 1,2); \quad \bar{\omega}_0 = \omega_0^{(0)} + \Delta_0.$$
$$(5.1.25)$$

In (5.1.25) $J_k(X)$ is the Bessel function; the shift of the level r ($r = 0, 1, 2$) under the influence of the faraway levels $k (k \neq 1, 2$ for $r = 1, 2)$ is approximately taken into account by means of $\Delta_r = 2^{-1} E_0 \times \sum_k \langle \varphi_k | z | \varphi_r \rangle^2 / \omega_{rk}^{(0)}$. Formulas (5.1.24) show that in the spectrum of any component $I_{m'',m'}^{(e)}(\Delta\omega)$ belonging to the allowed SLs, there is a central satellite (at the frequency $\Delta\omega = \bar{\omega}_2(m'') - \bar{\omega}_0(m') \equiv \Delta\omega_{20}$ and even satellites at the frequencies $\Delta\omega = \Delta\omega_{20} + 2s\omega$. Since $|\varepsilon_4/2| \ll |\varepsilon_2^{(2)}|$, the spectrum $I_{m'',m'}^{(e)}(\Delta\omega)$ is approximately symmetric with respect to the axis $\Delta\omega = \Delta\omega_{20}$. The spectrum of any forbidden component $\tilde{I}_{m'',m'}^{(e)}(\Delta\omega)$ consists of a set of odd satellites existing at the frequencies $(2s+1)\omega$ with respect to the position $\Delta\omega = \bar{\omega}_1(m'') - \bar{\omega}_0(m') \equiv \Delta\omega_{10}$. It should be emphasized, however, that the spectrum $\tilde{I}_{m'',m'}^{(e)}(\Delta\omega)$ is considerably *asymmetric* with respect to the axis $\Delta\omega = \Delta\omega_{10}$. This can be seen very strikingly in the multisatellite case $|\varepsilon_2^{(1)}/4\omega| \gg 1$. Indeed, the satellite intensity in the spectrum $\tilde{I}_{m'',m'}^{(e)}(\Delta\omega)$ is proportional to the value $g_s = (G_{s,1} + G_{s+1,1})^2$. Due to $G_{s,1} \approx J_s(\varepsilon_2^{(1)}/4\omega)$, $J_s(-x) = (-1)^s J_s(x)$, and $J_{s+1}(-x) = -(-1)^s J_{s+1}(x)$, the value of g_s (as a function of s) reaches its maximum at $s \approx \varepsilon_2^{(1)}/(4\omega)$. But if $s \approx -\varepsilon_2^{(1)}/4\omega$, then the terms $G_{s,1}$ and $G_{s+1,1}$, whose absolute values are close to one another, have different signs, and therefore considerably compensate each other in the expression for g_s.

The results may be used in spectroscopic measurements of strong microwave EFs (of $\sim$ 50–300 kV/cm) in spectra corresponding to the transitions $(3^1D, 3^1P) \rightarrow 2^1S, (3^1D, 3^1P) \rightarrow 2^1P$ of helium. To measure

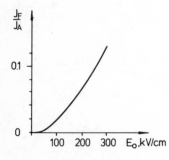

Fig. 5.6. The ratio of the total intensity of all satellites of the forbidden He I 5042 Å $(2^1S\text{-}3^1D)$ to the allowed He I 5016 Å $(2^1S\text{-}3^1P)$ lines vs. QEF amplitude E_0. The values are calculated for the spectrum polarization $e \parallel E_0$ and are relevant to QEF frequencies $\omega/2\pi \lesssim 600$ GHz

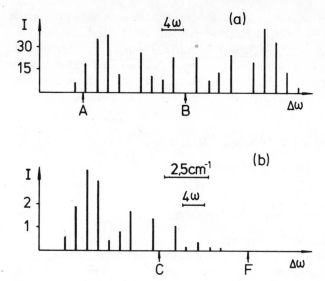

Fig. 5.7. Spectra of allowed (**a**) He I 5016 Å and forbidden (**b**) He I 5042 Å lines under the action of a QEF $E_0 \cos \omega t$ with $E_0 = 150$ kV/cm, $\omega/2\pi = 9.4$ GHz (polarization $e \parallel E_0$). The letter A marks the unperturbed position of the allowed line, B – the position $\Delta\omega = \Delta\omega_{20}$, C – the position $\Delta\omega = \Delta\omega_{10}$, F – the unperturbed position of the forbidden line

these fields the dependence on E_0 of the ratio of the total intensity of all satellites of the forbidden (J_F) and the allowed (J_A) SLs may be used (such a dependence is shown in Fig. 5.6 for the transition $(3^1D, 3^1P) \rightarrow 2^1S$, $e \parallel E_0$, $\omega/(2\pi) \leqslant 600$ GHz). In addition the profiles of the envelope of the satellites of the forbidden and allowed SLs can be used. Figure 5.7 shows the emission spectra ($e \parallel E_0$) in the vicinity of the allowed and the forbidden SLs for the transition $(3^1D, 3^1P) \rightarrow 2^1S$ in the QEF $E_0 \cos \omega t$ with the parameters $E_0 = 150$ kV/cm, $\omega/2\pi = 9.4$ GHz.

5.1.5 Modification of Helium-like Ion Satellites Caused by Mixing of Singlet and Triplet Terms

In high-temperature plasmas, when analyzing satellites of forbidden SLs of helium-like ions with large nuclear charges Z it is necessary to allow for the magnetic interactions of electrons (spin–orbit, spin–spin). Consider the three-level system of the levels of helium-like ions $0 \leftrightarrow 1^1S_0$, $1 \leftrightarrow 2^1S_0$, $2 \leftrightarrow 2^1P_1$. Due to the magnetic interactions, intensities of forbidden SL $1^1S_0 \leftrightarrow 2^1S_0$ are influenced not only by level 2, but also by level 3 $\leftrightarrow 2^3P_1$, since the WFs ψ_1, ψ_2, ψ_3 of levels 1, 2, 3 may be represented as an expansion in terms of the functions φ_1, φ_2, φ_3 of LS coupling [5.18]

$$\psi_1 = \varphi_1, \quad \psi_2 = (1 + \beta^2)^{-1/2}(\varphi_2 + \beta\varphi_3), \quad \psi_3 = (1 + \beta^2)^{-1/2}(\varphi_3 - \beta\varphi_2),$$
$$(5.1.26)$$

where β is a mixing coefficient. The allowance for $\beta \neq 0$ leads to a coupling of level 1 by dipole matrix elements both with level 2 and with level 3: $\langle \psi_1 | z | \psi_2 \rangle = z_{12}(1 + \beta^2)^{-1/2}$, $\langle \psi_1 | z | \psi_3 \rangle = -z_{12}\beta(1 + \beta^2)^{-1/2}$. Therefore at $\beta \neq 0$ the intensity of the allowed SL $2 \to 0$ decreases proportionally to $(1 + \beta^2)^{-1}$ and the intensities of forbidden line satellites are modified: they are determined by the values $|C_+(\beta)|^2$, $|C_-(\beta)|^2$, where the coefficients $C_+(\beta)$, $C_-(\beta)$ have the form

$$C_{\pm}(\beta) = -[2(1 + \beta^2)]^{-1} z_{12} E_0 [(\omega_{21}^{(0)} \pm \omega)^{-1} + (\omega_{31}^{(0)} \pm \omega)^{-1}\beta^2]. \quad (5.1.27)$$

This formula is valid under the conditions

$$|z_{12} E_0 [2(1 + \beta^2)^{1/2}(\omega_{21} \pm \omega)]^{-1}| \ll 1,$$

$$|\beta z_{12} [2(1 + \beta^2)^{1/2}(\omega_{31} \pm \omega)]^{-1}| \ll 1, \quad (5.1.28)$$

which at $\beta < 1$ are stronger than the inequality $|C_{\pm}| \ll 1$.

We emphasize that allowance for a coupling of sublevels 2^1S_0 and 2^3P_1 may significantly change satellite intensities even if $\beta^2 \ll 1$. This is related to the fact that for a wide range of nuclear charges Z, level 3 is located much more closely to level 1 than level 2:

	C V	O VII	Al XII	S XV	Ca XIX	Ti XXI	Fe XXV
$10^3\beta^2$	$3.4 \cdot 10^{-2}$	0.17	2.7	8.7	28	45	93
$10^3\omega_{31}^{(0)}/\omega_{21}^{(0)}$ ..	-3.5	-52	-89	-79	-49	-35	-11

For example, for the Ca XIX ion at $\omega/|\omega_{31}^{(0)}| = 5$, with the help of (5.1.27) it is easy to obtain that the allowance for a small mixing coefficient ($\beta^2 \approx 2.8 \times 10^{-2}$) modifies the ratio of satellite intensities by 60% ($|C_+(\beta)/C_-(\beta)|^2 |C_+(0)/C_-(0)|^{-2} \approx 1.6$).

We also analyze the case of Z so large that the contribution of level 2 to the intensity of forbidden line satellites is negligible compared to the contribution of level 3 and with a QEF frequency ω that obeys the inequality $|\omega_{31}| \ll \omega \ll |\omega_{21}|$. Then, neglecting ω_{31} compared to ω, the following expression for the perturbed WF of the level 1 may be obtained:

$$\psi_1 \approx \left\{ \varphi_1 \cos(\varepsilon \sin \omega t) - \varphi_2 \beta(1 + \beta^2)^{-1/2} \sum_{p=-\infty}^{+\infty} J_{2p-1}(\varepsilon) \exp[i(2p - 1)\omega t] \right\}$$

$$\times \exp(-i\omega_1 t), \quad \varepsilon \equiv z_{12} E_0 \beta[(1 + \beta^2)^{1/2}\omega]^{-1}. \quad (5.1.29)$$

The neglect of the contribution of level 2 to the intensity of satellites at frequencies $\pm\omega$ is justified under the condition $|z_{12} E_0/[2\omega_{21}(1 + \beta^2)^{1/2}]| \ll |\beta J_1(\varepsilon)|$.

From (5.1.29) it is seen that the intensities of forbidden line satellites are proportional to the squares of Bessel functions $J_{2p-1}^2(\varepsilon)$. Analogously, intensities of satellites for the allowed line ($2 \to 0$) are proportional to $J_{2p}^2(\varepsilon)$. In the considered case the QEF amplitude E_0 may be measured by the ratio of intensities $J_k^2(\varepsilon)/J_{k'}^2(\varepsilon)$ for satellites with different number k and k'. Such measurements may also be carried out in situations to which the method by Baranger and Mozer [5.1] is inapplicable, i.e., when the relative populations of the levels 1, 2, 3 are unknown or the volume in which the QEF exists does not coincide with the volume from which the allowed SL is emitted.

5.1.6 Satellites in a Stochastic QEF

Consider an EF of the type

$$E(t) = \sum_{j=1}^{J} E_j(t) \cos[\omega t + \varphi_j(t)], \qquad (5.1.30)$$

where $\varphi_j(t)$ and $E_j(t)$ are the random phases and amplitudes. More precisely, the phase and the amplitude experience a kangaroo-type process: their changes occur instantaneously (and simultaneously), the interval δt between these changes being a random quantity with a distribution density $\gamma \exp(-\gamma \delta t)$.

The influence of EF (5.1.30) on hydrogen SLs in plasmas (with an allowance for quasistatic EF F caused by ions) was first analyzed in [5.19]. In that paper it was shown that EF (5.1.30) leads to additional impact-like broadening $\Gamma_{\alpha\beta}$ and shift $D_{\alpha\beta}$ of each Stark component of a hydrogen SL; the values $\Gamma_{\alpha\beta}$ and $D_{\alpha\beta}$ were analytically calculated.

The problem of satellites of forbidden SLs of non-Coulomb radiators caused by the action of EF (5.1.30) was recently considered in [5.5] in the three-level scheme. The main results obtained are reviewed here.

In the limiting case $\gamma/\omega \to 0$ the satellites separated by $p\omega$ from a position of allowed ($p = 2k$; $k = 0, \pm 1, \ldots$) or forbidden ($p = 2l - 1$; $l = 0, \pm 1, \ldots$) SL have, generally speaking, an asymmetrical form, but with an increase of $|p|$ this asymmetry becomes less pronounced. For $|p| \gg 1$ the satellites have a Gaussian form with a characteristic width $\sim |p|^{1/2} b\omega$ and are additionally shifted by the value $|p|b\omega$, where $b \equiv (ez_{12}E_0/\hbar)^2/(2\omega\Delta)$, $E_0^2 = \sum_{j=1}^{J}\langle E_j^2(t)\rangle$.

In the opposite case $b \ll \gamma/\omega \ll 1$ each satellite has the Lorentz (dispersive) shape with the halfhalfwidth $|p|\gamma$. The central peak at the frequency of the allowed SL also has Lorentz shape, but its halfhalfwidth $\omega^2 b^2/2\gamma$ is significantly smaller than γ and decreases with the increase of γ (dynamic narrowing of the allowed SL).

5.2 Satellites of Dipole-Forbidden Spectral Lines in Resonant QEFs. Three-Level Scheme

5.2.1 Multiquantum Resonance in a Two-Level Subsystem

Consider the action of a QEF $E(t) = E_0 \cos \omega t$ a two-level subsystem (the levels 1 and 2 in Fig. 5.1). The treatment in the adiabatic basis (as in Sect. 5.1.2) leads to the system of equations (5.1.7) for amplitudes $C_1(t)$, $C_2(t)$ from (5.1.6).

We leave in the exponents in (5.1.7) only the terms with $\bar{\Delta}$ and ε_2, [defined in (5.1.10)]. Then the exponents may be represented in the form

$$\exp(\mp i\bar{\Delta}t) \sum_{m=-\infty}^{+\infty} J_m(A) \exp(\mp 2im\omega t), \qquad A \equiv \varepsilon_2/(2\omega). \tag{5.2.1}$$

If we first neglect nondiabatic effects, i.e., let $\dot{\beta}(t) \equiv 0$, then as solutions of the Schrödinger equation (5.1.3) we can take the time-periodic WFs

$$\chi_k(t) = \xi_k(t) \exp\left[-i \int_0^t dt'\omega_k(t')\right] \qquad (k = 1, 2), \tag{5.2.2}$$

which define two QSs. The separations between quasienergies of these QSs are equal to

$$Q = \bar{\Delta} + 2u\omega \quad (u = 0, \pm1, \pm2, \ldots). \tag{5.2.3}$$

In reality we have $\dot{\beta}(t) \not\equiv 0$, which may lead to nonadiabatic transitions between the QSs (5.2.2). The Fourier series expansion of $\dot{\beta}(t)$ has frequencies $(2v - 1)\omega$ only ($v = 0, \pm1, \pm2, \ldots$). It is clear that when $Q = (2v - 1)\omega$ a multiquantum resonance between many QS harmonics arises, caused simultaneously by *all* harmonics of $\dot{\beta}(t)$. Then the resonance condition can be finally written (introducing the detuning δ):

$$\bar{\Delta} = q\omega + \delta \quad (|\delta| \ll \omega), \tag{5.2.4}$$

where q is odd. However, further we shall consider (5.2.4) for even q also, which will be discussed below.

In the resonance approximation (5.2.4), leaving only slowly oscillating terms in (5.1.7), we obtain

$$\dot{C}_1 = -iC_2\Omega_0(q)\exp(-it\delta), \dot{C}_2 = -iC_1\Omega_0(q)\exp(it\delta), \tag{5.2.5}$$

where $\Omega_0(q)$ is a generalization of the Rabi frequency [5.20] (used often in nonlinear optics) for the case of a multiquantum multifrequency resonance (at $\delta = 0$)

$$\Omega_0(q) = (k\omega/8) \sum_{s=-\infty}^{+\infty} (a_{2s+2l-2} - a_{2s+2l})J_s(A), \quad q = 2l - 1,$$

$$\Omega_0(q) = 0, \quad q = 2l \tag{5.2.6}$$

($l = 1, 2, 3, \ldots$). Using the solution $C_1(t)$, $C_2(t)$ of the system (5.2.5), we find two WFs of QSs:

$$\psi_j(t) = \left\{ \exp[-i(\omega_1^{(0)} + \omega_2^{(0)})t/2 - i\lambda_j t - it\delta/2 + (i/2) \int_0^t dt'\,\omega_{21}(t')] \right\}$$

$$\times \left\{ \alpha_{1j}\chi_1(t) + \alpha_{2j}\chi_2(t)\exp\left[it\delta - i\int_0^t dt'\omega_{21}(t')\right] \right\}, \quad j = 1, 2;$$

$$\lambda_1 = \Omega_\delta, \quad \lambda_2 = -\Omega_\delta, \quad \Omega_\delta \equiv (\delta^2 + 4\Omega_0^2)^{1/2}/2,$$
$$\alpha_{11} = -\alpha_{22} = [(2\Omega_\delta - \delta)/4\Omega_\delta]^{1/2}, \quad \alpha_{12} = \alpha_{21} = [(2\Omega_\delta + \delta)/4\Omega_\delta]^{1/2}.$$
$$(5.2.7)$$

The quasienergies of these QS are equal to

$$\varepsilon_{1,2}(q) = (\omega_1^{(0)} + \omega_2^{(0)})/2 + q\omega/2 \pm [\delta^2 + 4\Omega_0^2(q)]/2. \qquad (5.2.8)$$

From (5.2.7,8) it becomes clear, in particular, that even-quantum resonances ($q = 2, 4, \ldots$) acquire physical meaning only with an allowance for a detuning δ: for them the Rabi frequency is $\Omega_\delta = \delta/2$. In practice the generalization of (5.2.4) to the case of even-quantum resonances allows us to carry out analytic calculations for a wider range of parameters.

5.2.2 Spectrum of Spontaneous Transitions to a Third Level in a Multiquantum Resonance

Spontaneous emission spectrum of a radiative transition from the two-level subsystem 1,2 (under conditions of multiquantum resonance) to the level 0 (Fig. 5.1) may be calculated with the help of WF (5.2.7) according to the general formula (1.3). This calculation gives

$$I^{(e)}(\Delta\omega) = \sum_{m_0, m_1, m_2} |r_{20e}|^2 J_0(A/2)$$
$$\times [\varepsilon_\pm (A_0 \mp \varepsilon_\mp B_1 \Omega_\delta/\Omega_0)^2 \delta(\Delta\omega \mp \omega/2 - \Omega_\delta)$$
$$+ \varepsilon_\mp (A_0 \pm \varepsilon_\pm B_1 \Omega_\delta/\Omega_0)^2 \delta(\Delta\omega \mp \omega/2 + \Omega_\delta)$$
$$+ \varepsilon_\mp (B_1/2)^2 \delta(\Delta\omega \pm 3\omega/2 - \Omega_\delta)$$
$$+ \varepsilon_\pm (B_1/2)^2 \delta(\Delta\omega \pm 3\omega/2 + \Omega_\delta)], \quad q = 1; \qquad (5.2.9)$$

$$I^{(e)}(\Delta\omega) = \sum_{m_0, m_1, m_2} |r_{20e}|^2 J_0(A/2)[\varepsilon_\pm A_0^2 \delta(\Delta\omega \mp q\omega - \Omega_\delta)$$
$$+ \varepsilon_\mp A_0^2 \delta(\Delta\omega \mp q\omega/2 + \Omega$$
$$+ \varepsilon_\mp (B_1/2)^2 \delta(\Delta\omega \pm (q-2)\omega/2 - \Omega_\delta)$$
$$+ \varepsilon_\pm (B_1/2)^2 \delta(\Delta\omega \pm (q-2)\omega/2 + \Omega_\delta)$$
$$+ \varepsilon_\mp (B_1/2)^2 \delta(\Delta\omega \pm (q+2)\omega/2 - \Omega_\delta)$$
$$+ \varepsilon_\pm (B_1/2)^2 \delta(\Delta\omega \pm (q+2)\omega/2 + \Omega_\delta)], \quad q = 2, 3, \ldots \qquad (5.2.10)$$

In (5.2.9,10) the upper signs correspond to the case of $\omega_2^{(0)} > \omega_1^{(0)}$, the lower signs to the case of $\omega_1^{(0)} > \omega_2^{(0)}$ (recall that the transition $2 \to 0$ is allowed); $\Delta\omega$ is counted from the frequency $(\omega_1^{(0)} + \omega_2^{(0)})/2 - \omega_0^{(0)}$; $r_{20} \equiv \langle \psi_2 | \mathbf{r} | \psi_0 \rangle$; the values $\varepsilon_+, \varepsilon_-$ are equal to

$$\varepsilon_+ = (2\Omega_\delta + \delta)/4\Omega_\delta, \qquad \varepsilon_- = (2\Omega_\delta - \delta)/4\Omega_\delta. \qquad (5.2.11)$$

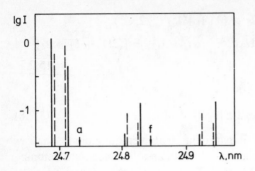

Fig. 5.8. The spectrum of $CV2^1S - (3^2P, 3^1D)$ lines in a three-quantum resonance with a CO_2-laser field ($E_0 = 40$ MV/cm). The *arrows* are the unperturbed positions of allowed (a) and forbidden (f) transitions. *Solid* lines– π–components, *dashed* lines– σ–components.

Summation over m_0, m_1, m_2 in (5.2.9,10) must be made, since in real radiators (helium or alkali atoms and corresponding ions) to the each of levels 0, 1, 2, several WFs, differing in their magnetic quantum numbers, may belong.

In (5.2.10) the terms in the first line are the splitting of an allowed SL, in the second line they are the splitting of a near satellite of a forbidden SL, and in the third line the splitting of a far satellite. In a single-quantum resonance the near satellite superimposes with the main line; therefore (5.2.9) has two terms less than (5.2.10).

As an example consider the SL of the C V ion, corresponding to the transitions $2^1S - (3^1P, 3^1D)$, in a field of a CO_2 laser. The distance $\Delta/2\pi c = 1846$ cm^{-1} between the unperturbed levels $3^1P, 3^1D$ is almost twice the frequency of the CO_2 laser $\omega/2\pi c \approx 943$ cm^{-1}. However, the laser field, moving the levels $3^1P, 3^1D$ apart, may tune them to the three-quantum resonance ($q = 3$). Such a resonance occurs in the range of $E_0 \approx (30–50)$ MV/cm. The effect manifests itself as the splitting $\Omega_\delta\lambda_0^2/\pi c$ of both the main SL and the satellites, which depends on E_0 (Fig. 5.8). Thus by measuring experimentally the distance $\Omega_\delta\lambda_0^2/\pi c$ between the sublines, it is possible to determine the QEF amplitude E_0 by (5.2.6–8). The value E_0 may be determined by measuring the subline intensities ratio and using (5.2.9–11). If the spectral resolution is insufficient, one can measure the separation of perturbed positions of the allowed SL $\bar\omega_a$ and the forbidden SL $\bar\omega_f = (\omega_S^+ + \omega_S^-)/2$ (ω_S^+, ω_S^- are the satellite frequencies averaged over the resonance splitting) and determine E_0 using the expression for $\bar\Delta$ from (5.1.10).

5.3 Satellites of Dipole-Forbidden Spectral Lines in More Complicated (Four-Level) Systems

5.3.1 QSs of a Three-Level Subsystem in a High-Frequency or Intense Field

We shall consider the action of a field $E(t) = E_0 \cos \omega t$ on a system with three levels (1,2,3) for which $z_{23} = 0$ (Oz $\parallel E_0$ configuration), $z_{12} \neq 0$ and $z_{13} \neq 0$

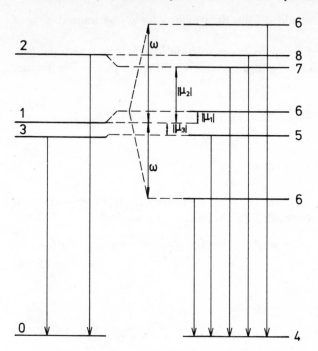

Fig. 5.9. Four unperturbed levels (of, e.g. a hydrogen-like ion) and corresponding quasienergy states in a strong field $E_0 \cos \omega t$. The *arrows* show radiative transitions. $0 - 1S_{1/2}$, $1 - 2S_{1/2}$, $2 - 2P_{3/2}$, $3 - 2P_{1/2}$, $4 - |1\ 0\ 1/2 \pm 1/2\rangle$, $5 - |2\ 1\ 1/2\ 1/2\rangle$, $6 - |2\ 0\ 1/2\ 1/2\rangle$, $7 - |2\ 1\ 3/2\ 1/2\rangle$, $8 - |2\ 1\ 3/2\ 3/2\rangle$)

(Fig. 5.9). The WFs will be found in the form

$$\psi(t) = \sum_{n=1}^{3} C_n(t)\varphi_n(t), \qquad \varphi_n(t) = \exp(-i\omega^{-1} E_0 \hat{z} \sin \omega t)\psi_n^{(0)}, \quad (5.3.1)$$

where $\psi_n^{(0)}$ represents the unperturbed WFs (we use atomic units with $\hbar = m = e = 1$). Substituting (5.3.1) into the Schrödinger equation, we obtain

$$\dot{C}_1 = -iC_1 V \sin^2 \alpha + C_2 \tilde{z}_{12} \sin \alpha (\omega_{23} \tilde{z}_{13}^2 + V \cos \alpha)$$
$$+ C_3 \tilde{z}_{13} \sin \alpha (-\omega_{23} \tilde{z}_{12}^2 + V \cos \alpha),$$
$$\dot{C}_2 = -C_1 \tilde{z}_{12} \sin \alpha (\omega_{23} \tilde{z}_{13}^2 + V \cos \alpha) - iC_2 (W \tilde{z}_{13}^2 + 2\tilde{z}_{12}^2 \tilde{z}_{13}^2 \omega_{23} \cos \alpha$$
$$+ V \tilde{z}_{12}^2 \cos^2 \alpha) - iC_3 \tilde{z}_{12} \tilde{z}_{13} [-W + (\tilde{z}_{13}^2 - \tilde{z}_{12}^2)\omega_{23} \cos \alpha + V \cos^2 \alpha],$$

$$(5.3.2)$$

where

$$\alpha \equiv B \sin \omega t, \quad B \equiv \xi E_0/\omega, \quad \xi \equiv (z_{12}^2 + z_{13}^2)^{1/2},$$
$$\tilde{z}_{12} \equiv z_{12}/\xi, \quad \tilde{z}_{13} \equiv z_{13}/\xi,$$
$$V \equiv \omega_{21} \tilde{z}_{12}^2 + \omega_{31} \tilde{z}_{13}^2, \qquad W \equiv \omega_{21} \tilde{z}_{13}^2 + \omega_{31} \tilde{z}_{12}^2.$$

The equation for C_3 is obtained from the equation for C_2 by transposition of the indices $2 \leftrightarrow 3$.

We shall now obtain an approximate solution in accordance with the general results from Sect. 2.3 or with the averaging principle [5.21]. The Fourier expansions

$$\sin\alpha = 2\sum_{k=1}^{\infty} J_{2k-1}(B)\,\sin(2k-1)\omega t, \quad \cos\alpha = J_0(B)$$

$$+ 2\sum_{k=1}^{\infty} J_{2k}(B)\,\cos 2k\omega t \qquad (5.3.3)$$

can be used to find the constant (a_{ij}) and oscillatory $(\tilde{a}_{ij}(t))$ components for each coefficient of the system (5.3.2). Omitting $\tilde{a}_{ij}(t)$, we find

$$\dot{C}_1 = -ia_{11}C_1, \quad \dot{C}_2 = -ia_{22}C_2 - ia_{23}C_3, \quad \dot{C}_3 = -ia_{32}C_2 - ia_{33}C_3,$$
$$(5.3.4)$$

where

$$a_{11} = [1 - J_0(2B)]V/2;$$
$$a_{23} = a_{32} = \tilde{z}_{12}\tilde{z}_{13}\{(\tilde{z}_{13}^2 - \tilde{z}_{12}^2)\omega_{23}[J_0(B) - 1] + [J_0(2B) - 1]V/2\};$$
$$a_{pp} = \omega_{p1} + (-1)^p 2\tilde{z}_{12}^2\tilde{z}_{13}^2\omega_{23}[J_0(B) - 1]$$
$$+ \tilde{z}_{1p}^2[J_0(2B) - 1]V/2, \quad (p = 2, 3). \qquad (5.3.5)$$

Solving the system (5.3.4) and substituting $C_n(t)$ into (5.3.1), we find three WFs of QSs of a three-level system in a field $\boldsymbol{E}(t)$:

$$\psi_1(t) = \varphi_1 \exp(-i\mu_1 t), \quad \psi_2(t) = (1 + A^2)^{-1/2}(\varphi_2 - A\varphi_3)\exp(-i\mu_2 t),$$
$$\psi_3(t) = (1 + A^2)^{-1/2}(A\varphi_2 + \varphi_3)\exp(-i\mu_3 t),$$
$$A \equiv (\mu_3 - a_{33})/a_{23} = (a_{22} - \mu_2)/a_{23}, \qquad (5.3.6)$$

where

$$\mu_1 = a_{11}, \quad \mu_{2,3} = (a_{22} + a_{33})/2 \pm [(a_{22} - a_{33})^2/4 + a_{23}^2]^{1/2} \qquad (5.3.7)$$

are the quasienergies (Fig. 5.9).

The limits of validity of this solution are set by the following conditions [subject to which we can ignore $\tilde{a}_{ij}(t)$]:

$$|VJ_{2k}(2B)|/2k\omega \ll 1,$$
$$|2(\tilde{z}_{13}^2 - \tilde{z}_{12}^2)\omega_{23}J_{2k}(B) + VJ_{2k}(2B)|\tilde{z}_{12}\tilde{z}_{13}/2k\omega \ll 1,$$
$$|2\omega_{23}\tilde{z}_{13}^2 J_{2k-1}(B) + VJ_{2k-1}(2B)|\tilde{z}_{12}/(2k-1)\omega \ll 1,$$
$$|4\tilde{z}_{13}^2\omega_{23}J_{2k}(B) + VJ_{2k}(2B)|\tilde{z}_{12}^2/2k\omega \ll 1, \qquad (5.3.8)$$

where $k = 1, 2, 3, \ldots)$, and also by two other conditions which are obtained from the system (5.3.8) by the index transposition $2 \leftrightarrow 3$. The stringent conditions of the system (5.3.8) can be simplified to a less accurate but physically clearer requirement:

$$\max((z_{12}^2 + z_{13}^2)^{1/2} E_0 \omega, \omega^2) \gg [(\omega_{21} z_{12}^2 + \omega_{31} z_{13}^2)/(z_{12}^2 + z_{13}^2)]^2. \quad (5.3.9)$$

It follows that the solution represented by (5.3.6,7) is valid either if the field is strong or if its frequency is high.

5.3.2 Radiative Transitions from the States $2P_{1/2}$, $2S_{1/2}$, $2P_{3/2}$ of a Hydrogen-like Ion in a High-Frequency or Intense Field

Consider the following states of a hydrogen-like ion, regarding them as an unperturbed three-level system: $1 : |2\ 0\ 1/2\ 1/2\rangle$, $2 : |2\ 1\ 3/2\ 1/2\rangle$, $3 : |2\ 1\ 1/2\ 1/2\ 1/2\rangle$ (Fig. 5.9). The WFs (5.3.6) can be used to find quite readily the spontaneous emission spectrum in a field $E(t)$ for transitions to the states $0^{(m)} : |1\ 0\ 1/2\ m\rangle$, where $m = \pm 1/2$:

$$
\begin{aligned}
I(\Delta\omega) = \sum_{m=-1/2}^{+1/2} \{ &(1 + A^2)^{-1} \{ |M_{23}^{(m)}|^2 \delta(\Delta\omega - \mu_2) + |M_{32}^{(m)}|^2 \delta(\Delta\omega - \mu_3) \\
&+ \sum_{\substack{k=-\infty \\ k \neq 0}}^{+\infty} J_{2k}^2(B) [|N_{23}^{(m)}|^2 \delta(\Delta\omega - \mu_2 - 2k\omega) \\
&+ |N_{32}^{(m)}|^2 \delta(\Delta\omega - \mu_3 - 2k\omega)]\} \\
&+ |R^{(m)}|^2 \sum_{k=-\infty}^{+\infty} J_{2k-1}^2(B) \delta(\Delta\omega - \mu_1 - (2k - 1)\omega)\};
\end{aligned}
$$

$$R^{(m)} = r_{20(m)} z_{12} + r_{30(m)} z_{13}, \qquad N_{23}^{(m)} = R^{(m)}(\tilde{z}_{12} - A\tilde{z}_{13}),$$

$$N_{32}^{(m)} = R^{(m)}(\tilde{z}_{13} + \tilde{A}z_{12}), \qquad M_{23}^{(m)} = J_0(B) N_{23}^{(m)} - Q_{23}^{(m)}(A\tilde{z}_{12} + \tilde{z}_{13}),$$

$$M_{32}^{(m)} = J_0(B) N_{32}^{(m)} - Q_{32}^{(m)}(\tilde{z}_{12} - A\tilde{z}_{13}), \qquad R^{(1/2)} = R_{10}^{21} e_z/\sqrt{3},$$

$$Q_{23}^{(-1/2)} = -Q_{32}^{(-1/2)} = R_{10}^{21}(e_x - i e_y)/\sqrt{6},$$

$$R^{(-1/2)} = Q_{23}^{(1/2)} = Q_{32}^{(1/2)} = 0, \qquad z_{12} = (2/3)^{1/2}, \qquad z_{13} = (1/3)^{1/2}. \tag{5.3.10}$$

(R_{10}^{21} is a radial integral [5.22], $\Delta\omega$ is relative to ω_{10}). From (5.3.10) it is seen that odd satellites of the forbidden SL ($1S_{1/2} - 2S_{1/2}$) and even satellites of the allowed SL ($1S_{1/2} - 2S_{1/2}, 2P_{3/2}$) arise. In addition there is also an unshifted SL $\Delta\omega = \omega_{21}$ with an intensity $(R_{10}^{21})^{2/3}$, not shown in (5.3.10). In the limit $\omega_{21} \to 0$, $\omega_{31} \to 0$ the results coincide with that of *Blochinzew* [5.23].

Figure 5.10 shows part of the calculated spectrum of the L_α line of the O VIII ion in a field of a neodymium laser (the power density reaching the target was $P \approx 4.6 \times 10^{14}$ W/cm², i.e. the field intensity was $E_0 \approx 0.59$ GV/cm), including the adjacent satellite region. The appearance of such laser satellites

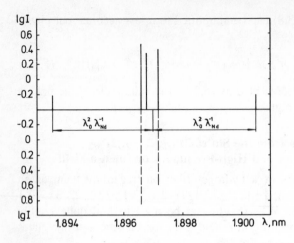

Fig. 5.10. Spectrum of the L_α line of the O VIII ion in a field of a Nd laser with $E_0 = 0.59$ GV/cm. Three main components (corresponding to transitions from the states $|2\ 1\ 3/2\ 3/2\rangle$, $|2\ 1\ 3/2\ 1/2\rangle$, $|2\ 1\ 1/2\ 1/2\rangle$ in Fig. 5.9) and the nearest satellites are shown. The *dashed lines* are the unperturbed intensities of the doublet (shown by reflection in the abscissa)

(located well outside the dielectronic satellite range) may provide a diagnostic method for high-temperature plasmas as powerful as the use of the Baranger and Mozer helium satellites [5.1] employed in studies of low-temperature plasmas. A local amplitude E_0 in a plasma corona can be determined by two methods: from the ratio of the intensities of the satellites, [see (5.3.10)], and from the shift of spectral components identical with the quasienergies μ_{1-3}, [see (5.3.7, 5)].

5.4 Electron Oscillatory Shift in Plasmas Interacting with a Powerful Coherent Radiation

5.4.1 Calculation in the Rectilinear Trajectories Approximation

In all modern versions of the theory of SL impact broadening by plasma electrons, the nonzero effects of line width and shift arise only in the second order of the Dyson expansion of the time-evolution operator of a radiating particle – the first-order terms cancel after averaging over the isotropic Maxwell distribution of electron velocities. However, in a plasma interacting with laser radiation the electron velocity distribution acquires anisotropic corrections. Therefore even in the first order of the Dyson expansion a nonzero effect arises: a shift of the SL that we have called the "electron oscillatory shift" (EOS). The dominant contribution to EOS is the quadrupole interaction of a radiator with perturbing electrons.

Consider a radiating particle with one electron outside the completed shells. The potential of its interaction with a free plasma electron may be given by the known multipole expansion [Ref. 5.24, Eg. (58)]:

$$U(t) = e^2 \left[\frac{\boldsymbol{r}\boldsymbol{r}_e}{|\boldsymbol{r}_e|^3} + \frac{3}{2} \left(\frac{(\boldsymbol{r}\boldsymbol{r}_e)(\boldsymbol{r}\boldsymbol{r}_e)}{|\boldsymbol{r}_e|^5} - \frac{r^2}{|\boldsymbol{r}_e|^3} \right) + \cdots \right]. \tag{5.4.1}$$

Here $r_e(t)$ is the radius-vector of a free electron, r is the radius-vector operator of an optical electron of a radiator (all measured from the nucleus of the radiator); it is assumed that $r_e > r$.

We use the rectilinear trajectories approximation (RTA) of perturbing electrons

$$r_e(t) = \rho + Vt, \qquad \rho V = 0, \tag{5.4.2}$$

counting time from the point of closest approach. The substitution of (5.4.2) in (5.4.1) and integration over t brings the following result for the first order terms in the Dyson expansion of the radiator time-evolution operator:

$$S(\rho, V) - 1 = -(\mathrm{i}/\hbar) \int_{-\infty}^{+\infty} U(t)\,\mathrm{d}t$$

$$= -\frac{\mathrm{i}\hbar}{m_e \rho V} \left\{ 2\frac{r\rho}{a_0\rho} + \frac{a_0}{\rho} \left[2\left(\frac{r\rho}{a_0\rho}\right)^2 + \left(\frac{rV}{a_0 V}\right)^2 - \frac{r^2}{a_0^2} \right] \right\}, \tag{5.4.3}$$

where $a_0 \equiv \hbar^2/m_e e^2$ is the Bohr radius.[2] Averaging over an angle $\theta' = (\rho, r)$ gives

$$S(\rho, V) - 1 = -\left(\frac{\mathrm{i}\hbar}{m_e \rho^2 a_0} \right) \left[\frac{(rV)^2}{V^3} - \frac{r^2}{3V} \right]. \tag{5.4.4}$$

The impact width γ_e and shift d_e are usually expressed through an electron impact broadening operator Φ_e:

$$\gamma_e + \mathrm{i}d_e = -\Phi_e = N_e \int\!\!\int\!\!\int V f(V)\,\mathrm{d}V \int_0^\infty 2\pi\rho[1 - S(\rho, V)]\,\mathrm{d}\rho. \tag{5.4.5}$$

In the case of an isotropic velocity distribution $f(V)$ averaging over the directions of the vector V leads to the known result $\Phi_e = 0$ (in the first order of the Dyson expansion).

Calculation of the shift. Consider the nontrivial case in which relatively slow oscillations of electrons under the action of a QEF $E_0 \cos \omega t$ are imposed on their rapid passing by a radiator. In this case a perturbing electron velocity may be represented as $V = V_0 + v$. Here V_0 is characterized by the isotropic Maxwell distribution $f_M(V_0)$; the value $v \parallel E_0$ is distributed by the law $\pi^{-1}(2\tilde{v}^2 - v^2)^{-1/2}$, where $\tilde{v} = (eE_0/m_e\omega)^2/2$. Averaging over an angle θ between V_0 and v gives

$$\gamma_e + \mathrm{i}d_e = \mathrm{i}\frac{2\pi e^2}{3\hbar}\langle Q_{zz}\rangle G\left(\frac{\tilde{v}^2}{v_{Te}^2}\right) N_e \int \frac{\mathrm{d}\rho}{\rho};$$

---

[2] Note that in the analogous formula (105) from [5.24] the minus sign before the braces is incorrect.

$$G(y) \equiv \int\limits_{-\tilde{v}\sqrt{2}}^{\tilde{v}\sqrt{2}} \frac{\mathrm{d}v}{\pi(2\tilde{v}^2 - v^2)^{1/2}} \int\limits_{0}^{\infty} g\left(\frac{v}{V_0}\right) f_{\mathrm{M}}(V_0)\, \mathrm{d}V_0,$$

$$g(x) \equiv (5 - 3/x^2)/8 + [3(1 - x^2)^2/(32x^3)]\, \ln[(1 + x^2)/(1 - x^2)],$$

$$(5.4.6)$$

where $\langle Q_{zz}\rangle = \langle 3z^2 - r^2\rangle$ is the zz-component of a quadrupole moment tensor in the considered state of the radiator; $v_{\mathrm{Te}} \equiv (2T_e/m_e)^{1/2}$.

Thus arises the additional shift d_e of radiator energy levels and of the SL. The dominant contribution to this EOS effect comes from the quadrupole interaction of the radiator with perturbing electrons.

The function $g(x)$ may be simply approximated by the formula $g(x) \approx x^2/(x^2 + 5)$, which correctly describes the limiting cases $x \ll 1$ and $x \gg 1$ and allows us to obtain a very suitable approximation for $G(v^2/v_{\mathrm{Te}}^2)$:

$$G(y) = 2y/(y + 5), \quad y \leqslant 5, \qquad G(y) = 1, \quad y \geqslant 5. \qquad (5.4.7)$$

For the radiators with one electron outside completed shells the quadrupole moment is

$$\langle Q_{zz}\rangle = \left[1 - \frac{3m_j^2}{(j^2 + j)}\right][5n^2 + 1 - 3l(l + 1)]\frac{n^2 a_0^2}{4Z_{\mathrm{eff}}^2}, \qquad (5.4.8)$$

where Z_{eff} is the effective charge of the atomic core.

To obtain a finite value of the EOS it is necessary to cut off the integral over ρ in (5.4.6) at large and small impact parameters. In the case considered we have $\omega \geqslant \omega_{\mathrm{pe}} = (4\pi e^2 N_e/m_e)^{1/2}$, so that as an upper cutoff parameter the value $\rho_{\max} = v_{\mathrm{Te}}/\omega \leqslant \rho_{\mathrm{De}}$ (and not ρ_{De} as usually) should be taken. As a lower cutoff the characteristic size of an excited radiator $\rho_r = n^2 a_0/Z_{\mathrm{eff}}$ may be taken [recall that the multipole expansion (5.4.1) is valid at $\rho \gtrsim \rho_r$]. Then using (5.4.7, 8) the following expression for d_e may be easily obtained:

$$d_e = \left(\frac{2\pi e^2}{3\hbar}\right)\langle Q_{zz}\rangle G\left(\frac{\tilde{v}^2}{v_{\mathrm{Te}}^2}\right) N_e \ln\left(\frac{v_{\mathrm{Te}} Z_{\mathrm{eff}}}{n^2 a_0 \omega}\right). \qquad (5.4.9)$$

Allowance for higher multipoles. For this purpose, instead of the expansion (5.4.1) we use the exact expression for the potential

$$U(t) = e^2(|\mathbf{r}_e - \mathbf{r}|^{-1} - |\mathbf{r}_e|^{-1}). \qquad (5.4.10)$$

Substituting (5.4.2) into (5.4.10) and integrating over t we obtain

$$S(\rho, \mathbf{V}) - 1 = -\frac{i}{\hbar}\int\limits_{-\infty}^{+\infty} U(t),\, \mathrm{d}t$$

$$= i\frac{e^2}{\hbar V} \ln\left[1 + \left(-2\mathbf{r}\rho + r^2 - \frac{(\mathbf{r}V)^2}{V^2}\right)\rho^{-2}\right]. \qquad (5.4.11)$$

[Note that (5.4.3) may be reproduced from (5.4.11) by expansion in the small parameter $r/\rho \sim \rho_r/\rho$.] The result of averaging over the anisotropic distribution of the velocity $V = V_0 + v$ may be, by analogy, represented in the approximate form

$$d_e = (2\pi e^2/\hbar) N_e \int d\rho \, \rho \ln[1 + \langle Q_{zz} \rangle G(\tilde{v}^2/v_{Te}^2)/3\rho^2], \qquad (5.4.12)$$

which, upon expansion of the logarithm in a series in the small parameter $|\langle Q_{zz} \rangle|^{1/2}/\rho \sim \rho_r/\rho$, coincides exactly with the corresponding expression in (5.4.6).

It is essential that the integral in (5.4.12) does not diverge at small ρ [in contrast to the integral in (5.4.6)]. Therefore the cutoff at small ρ may not be introduced and correspondingly the ratio ρ_r/ρ should not be assumed to be small. Integrating over ρ from 0 to $\rho_{max} = v_{Te}/\omega \gg |\langle Q_{zz} \rangle|^{1/2}$, we obtain the final expression (in the RTA) for the value of the EOS:

$$d_e = \frac{\pi e^2}{3\hbar} \langle Q_{zz} \rangle G\left(\frac{\tilde{v}^2}{v_{Te}^2}\right) N_e \left(\ln \frac{6T_e}{m_e \omega^2 |\langle Q_{zz} \rangle| G(\tilde{v}^2/v_{Te}^2)} + 1\right). \qquad (5.4.13)$$

The first term in the final parentheses in (5.4.13) corresponds to the contribution of the quadrupole interaction, the second term allows for the contribution of all higher multipoles. We emphasize that the appearance of the logarithm in (5.4.9, 13) reflects the dominating role of distant electrons.

5.4.2 Calculations Including Curved Trajectories

Shift at fixed velocities. In the Coulomb field of a radiating ion a perturbing electron moves along a hyperbolic trajectory. After integrating over such a trajectory we obtain instead of (5.4.4) the expression

$$S(\rho, V) - 1 = -i\frac{\hbar}{m_e \rho^2 a_0}\left(\frac{(rV)^2}{V^3} - \frac{r^2}{3V}\right) C(\varepsilon);$$

$$C(\varepsilon) \equiv 1 + 3/(2\varepsilon^2) - \varepsilon^{-4} + 3[\pi - \arccos(\varepsilon^{-1})]/[2(\varepsilon^2 - 1)^{1/2}], \qquad (5.4.14)$$

where ε is the eccentricity of the trajectory.

The further steps are analogous to the RTA: the substitution of (5.4.14) into (5.4.5), the representation $V = V_0 + v$, the averaging over the angle $\theta = (V_0, v)$ [the integration over $\cos \theta$ may be reduced to a simpler integration over V^2 using $\cos \theta = (V^2 - V_0^2 - v^2)/2V_0 v$]. Then multiplying by $2\pi \rho N_e$ and integrating over ρ we obtain

$$2\pi N_e \int\limits_{\rho_{min}}^{\rho_{max}} \langle V[1 - S(\rho, V)]\rangle_\theta \rho \, d\rho = i\frac{2\pi \hbar \langle Q_{zz} \rangle N_e}{3m_e \alpha_0} I_1;$$

$$I_1(v \ll V_0) \approx \frac{1}{10}\left[\frac{\pi - \arctan\eta}{\eta} + \ln\frac{\eta^3}{(1+\eta^2)^{1/2}} + \frac{1}{12}\left(-\frac{48}{1+\eta^2}\right.\right.$$

$$\left.\left. + \frac{166}{(1+\eta^2)^2} - \frac{124}{(1+\eta^2)^3} - \frac{5/2}{(1+\eta^2)^4}\right)\right]\left(\frac{v}{V_0}\right)^2\Bigg|_{\eta_{\min}}^{\eta_{\max}},$$

$$I_1(v \gg V_0) \approx \left[\ln\eta - \left(ln\eta + \frac{3\pi}{4\eta}\right)\frac{V_0^2}{v^2}\right]\Bigg|_{\eta_{\min}}^{\eta_{\max}}, \tag{5.4.15}$$

where

$$\eta_{\max} = m_e V_0^2 \rho_{\max}/[(Z_{\text{eff}} - 1)e^2], \qquad \eta_{\min} = m_e V_0^2 \rho_{\min}/[(Z_{\text{eff}} - 1)e^2], \tag{5.4.16}$$

The cutoff parameters should be chosen just as in the case of the RTA: $\rho_{\max} = v_{\text{Te}}/\omega$, $\rho_{\min} = n^2\alpha_0/Z_{\text{eff}}$.

Averaging over velocities. The expression for the value of the EOS $d_e(V_0, v)$ at fixed V_0 and v is obtained from (5.4.15) by removing the imaginary unity. We average $d_e(V_0, v)$ over the Maxwell distribution $f_M(V_0)$:

$$\langle d_e(V_0, v)\rangle_{V_0} \equiv d_e(v) = [2\pi\hbar\langle Q_{zz}\rangle N_e/3m_e\alpha_0]I_2;$$

$$I_2 \equiv \int\limits_{V_{\min}}^{\infty} I_1(\eta(V_0))V_0^2 v_{\text{Te}}^{-3}\exp(-V_0^2/v_{\text{Te}}^2)\,dV_0, \tag{5.4.17}$$

We need to cut off this integral at $V_{\min}$ since it diverges at the lower limit. The divergence [caused by the first term in $I_1(v \ll V_0)$ from (5.4.15)] physically means that for hyperbolic (nonrectilinear) trajectories, the dominant contribution to the EOS comes from slow electrons.

The value $V_{\min}$ should be chosen from the condition of validity of electron impact broadening, which follows from the second order of the Dyson expansion of the radiator evolution operator. Then we find that

$$V_{\min} = n^2\hbar/m_e Z_{\text{eff}}\rho_{\min} = e^2/\hbar \equiv v_B \tag{5.4.18}$$

coincides with the electron velocity in the first Bohr orbit in a hydrogen atom.

Substituting $I_1(\eta(V_0))$ from (5.4.16) into (5.4.17) and integrating over V_0 we obtain the final expression for the value of the EOS. However, it is rather cumbersome [5.7] and we do not reproduce it here.

5.4.3 Discussion

The typical form of the dependence of d_e on $\tilde{v}/v_{\text{Te}}$ is shown in Fig. 5.11. It is seen that the allowance for curvature in the perturbing electron trajectories leads to two qualitatively new results that do not appear in the RTA: 1) the sign-reversal in d_e vs $\tilde{v}/v_{\text{Te}}$; 2) the linear dependence $d_e \propto \tilde{v} \propto E_0$ at small $\tilde{v}/v_{\text{Te}}$, since at $\tilde{v} \ll v_{\text{Te}}$ the integral in (5.4.17) is $I_2 \approx [\pi(Z_{\text{eff}} - 1)\rho_r/10n^2\rho_{\max}](v_B/v_{\text{Te}})^2 v/v_{\text{Te}}$. This linear dependence upon the laser field

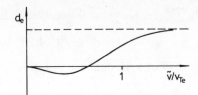

Fig. 5.11. Dependence of an electron oscillatory shift of a spectral line component on the ratio of the oscillatory velocity of the electron $\tilde{v} = eE_0/2^{1/2}m_e\omega$ in a laser field to its thermal velocity $v_{Te} = (2T_e/m_e)^{1/2}$

seems to be the most remarkable feature of the EOS effect. Recall that the direct action of a laser field on radiators leads to a weaker shift $d_E \propto E_0^2$ [5.18].

The conditions for observation of the EOS might be most favorable for lithium-like ions. An example of this is the SL of C IV at 38.418 nm ($2P_{3/2} - 3D_{5/2}$) in a dense plasma irradiated by the second harmonic of an Nd laser. At the power density $q = 5 \times 10^{13}$ W/cm^2 and plasma parameters $N_e = 2 \times 10^{21}$ cm^{-3}, $T_e = 4 \times 10^5$ K, for the σ-component of this line the EOS is $\Delta\lambda = d_e\lambda_0^2/2\pi c = 0.011$ nm (and corresponds to the negative values of d_e in Fig. 5.11). An example in a low-temperature plasma is the SL of Be II at 151.241 nm ($2P_{3/2} - 3D_{5/2}$) under a CO$_2$ laser radiation. At the power density of $q = 1.4 \times 10^{10}$ W/cm^2 and plasma parameters $N_e = 5 \times 10^{18}$ cm^{-3}, $T_e = 4 \times 10^4$ K for the σ-component of this line the EOS is equal to $\Delta\lambda \approx 0.017$ nm.

In both examples the obtained red shifts exceed the estimations which allow only for d_E and an ion quadrupole static shift [5.18]. Therefore in the anomalous shifts of SLs in laser plasmas [5.25] the EOS effect does appear to play an important role.

Note in conclusion that for calculations of the EOS for any SL as a whole, it is necessary to multiply the shift difference of upper (m_j) and lower (m_j') magnetic sublevels by squares of matrix elements (which determine intensities of SL components) and to sum over all components. One can derive a useful relation connecting the shifts of the same SL detected along ($\overline{\Delta d_{\parallel}}$) and transverse ($\overline{\Delta d_{\perp}}$) to the laser field E_0.

Indeed, from (5.4.9,17) it is seen that the dependence of d_e on m_j is determined by the dependence of the quadrupole moment $\langle Q_{zz} \rangle \equiv Q(m_j) \propto [1 - 3m_j^2/(j^2 + j)]$, see (5.4.8). Therefore the quantities $\overline{\Delta d_{\parallel}}$ and $\overline{\Delta d_{\perp}}$ may be represented

$$\overline{\Delta d_{\parallel}} = k \sum_{m_j, m_j'} [Q(m_j) - Q(m_j')] \left(\left| x_{bm_j'}^{am_j} \right|^2 + \left| y_{bm_j'}^{am_j} \right|^2 \right),$$

$$\overline{\Delta d_{\perp}} = k \sum_{m_j, m_j'} [Q(m_j) - Q(m_j')] \left(\left| x_{bm_j'}^{am_j} \right|^2 + \left| z_{bm_j'}^{am_j} \right|^2 \right), \qquad (5.4.19)$$

where k does not depend on m_j, m_j'; a and b are totalities of other (non-magnetic) quantum numbers. It is known that $\sum_{m_j'} |r_{bm_j'}^{am_j}|^2$ does not depend on

m_j [5.22]. Using this property and the equality $\sum_{m=-j}^{j} Q(m_j) = 0$ we obtain

$$\overline{\Delta d_{\parallel}} = -2\overline{\Delta d_{\perp}}. \tag{5.4.20}$$

Equation (5.4.20) is a universal relation for the EOS of a SL due to a transition between any two multiplets.

5.5 Action of QEFs on Diatomic Polar Molecules

5.5.1 Satellites in Vibrational-Rotational Spectra

Let us consider electric dipole transitions in a diatomic polar molecule (DPM) between two electronic states (Fig. 5.12). Assume that the upper electronic term a and the lower electronic term b are singlet terms, the absolute value of the electronic orbital momentum projection onto the molecular axis for term a being equal to Λ_a and for term $b - \Lambda_b$. Suppose that initially the molecule is in the upper state $|\Lambda_a, v_a, J\rangle$ (v and J are the vibrational and rotational quantum numbers, respectively). Then in the absence of the external EF, when $\Lambda_a - \Lambda_b = 0, \pm 1$ the transitions $|\Lambda_a, v_a, J\rangle \rightarrow |\Lambda_b, v_b, J'\rangle$ are allowed, where $J' = J \pm 1, J$ (for $J' = J$ the condition $\Lambda_a^2 + \Lambda_b^2 \neq 0$ is an additional requirement).

Let an external static EF F act on the molecule along the Oz axis. The Hamiltonian of the molecule in the field F is $H = H_0 - \mu_z F$, where H_0 is the unperturbed Hamiltonian and μ_z the projection of the molecular dipole moment onto the axis Oz. Then, using first-order perturbation theory we obtain the following: the state of each rotational level $|\Lambda, v, J\rangle$ of the molecule acquires an admixture of states $|\Lambda, v, J \pm 1\rangle$, the amount of admixture being proportional to the value of F. We consider Molecular Emission Spectra (MES), assuming that e is a unit vector of polarization of the emitted photons. Then the SL intensity corresponding to the transition $|\Lambda_a, v_a, J\rangle \rightarrow |\Lambda_b, v_b, J'\rangle$ is proportional to the quantity $|\langle \Lambda_a, v_a, J | \mu e | \Lambda_b, v_b, J'\rangle|^2$. Therefore at $F \neq 0$

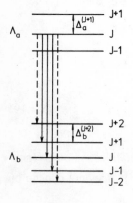

Fig. 5.12. The scheme of diatomic polar molecule energy levels. *Solid arrows* indicate allowed transitions. *Dashed arrows* show transitions forbidden in the absence of an external EF

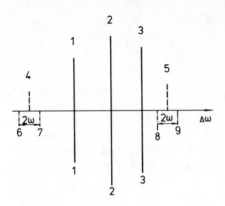

Fig. 5.13. Spontaneous emission spectrum of a polar molecule in a static EF F (above the abscissa) or in a QEF $E_0 \cos \omega t$ (below the axis). 1–3 are allowed components [1 : $J - (J + 1)$, 2 : $J - J$, 3 : $J - (J - 1)$]; 4, 5 are forbidden components [4 : $J - (J + 2)$, 5 : $J - (J - 2)$]; 6–9 are forbidden components satellites [6 : $S^+_{J-(J+2)}$, 7 : $S^-_{J-(J+2)}$, 8 : $S^-_{J-(J-2)}$, 9 : $S^+_{J-(J-2)}$]

in MES, besides allowed lines (whose intensities in this approximation are practically independent of F), forbidden SLs appear, corresponding to the transitions $|\Lambda_a, v_a, J\rangle \rightarrow |\Lambda_b, v_b, J \pm 2\rangle$ (Fig. 5.13). Their intensities are proportional to F^2. We calculate the intensities assuming that the vector e constitutes an angle θ to the direction of Oz. Then allowed SL intensities $|\Lambda_a, v_a, J\rangle \rightarrow |\Lambda_b, v_b, J \pm 1\rangle$ can be represented as

$$I_{J \to (J \pm 1)} = C_{ab} N_J P^2(\Lambda_a, J; \Lambda_b, J \pm 1) \tag{5.5.1}$$

and for forbidden SL intensities $|\Lambda_a, v_a, J\rangle \rightarrow |\Lambda_b, v_b, J \pm 2\rangle$ we have

$$I_{J \to (J+k)} = \frac{C_{ab} N_J F^2 (3 + \cos^2 \theta)}{10(2J + 1 + k)\hbar^2} \left[P(\Lambda_a, J; \Lambda_b, J + k/2) \right.$$

$$\times \left(\frac{(J + n_1)^2 - \Lambda_b^2}{J + n_1} \right)^{1/2} \frac{\mu_b}{\Delta_b^{(J+n_1)}}$$

$$\left. - P(\Lambda_a, J+k/2; \Lambda_b, J+k) \left(\frac{(J+n_2)^2 - \Lambda_a^2}{J+n_2} \right)^2 \frac{\mu_a}{\Delta_a^{(J+n_2)}} \right]^2, \tag{5.5.2}$$

where $k = \pm 2$; $n_1(k = 2) = 2$, $n_2(k = 2) = 1$, $n_1(k = -2) = -1$, $n_2(k = -2) = 0$. In (5.5.1,2) N_J is the population of the level $|\Lambda_a, v_a, J\rangle$, the coefficient C_{ab} is independent of J, and the values P are

$$P(\Lambda, J; \Lambda, J - 1) = J^{-1/2}(J^2 - \Lambda^2)^{1/2},$$

$$P(\Lambda - 1, J; \Lambda, J - 1) = (4J)^{-1/2}[(J - \Lambda)(J - \Lambda + 1)]^{1/2},$$

$$P(\Lambda, J; \Lambda - 1, J - 1) = (4J)^{-1/2}[(J + \Lambda)(J + \Lambda - 1)]^{1/2},$$

$$P(\Lambda_1, J_1; \Lambda_2, J_2) = P(\Lambda_2, J_2; \Lambda_1, J_1). \tag{5.5.3}$$

In (5.5.2), $\Delta_v^{(J)} = 2J B_v$ (B_v is the rotational constant for the state v); μ_v is the dipole moment of the molecule in the state v ($v = a, b$). Equations (5.5.2) are valid if the following conditions are satisfied

$$(2\hbar B_\nu J_\nu)^{-1}(4J_\nu^2 - 1)^{-1/2}(J_\nu^2 - \Lambda_\nu^2)^{1/2}\mu_\nu F \ll 1, \tag{5.5.4}$$

$$[2\hbar B_\nu J_\nu(J_\nu + 1)]^{-1}\Lambda_\nu\mu_\nu F \ll 1,$$

$$\nu = a, b; \quad J_a = J, J+1; \quad J_b = J-2, J-1, \ldots, J+2. \tag{5.5.5}$$

If (5.5.4) is satisfied, the interaction with the field F is small compared to the separation of the close rotational levels of the state ν ($\nu = a, b$). Condition (5.5.5) means that the repulsion in the field F of the Λ-doublet sublevels of the rotational levels J_ν is small compared to the separation of the neighboring levels $J_\nu + 1$. Formulas (5.5.1,2) have been obtained assuming equal populations in states with different M belonging to the upper level J (at $\Lambda_a = 0$) or to each of the Λ - doublets of the level J (at $\Lambda_a \geqslant 1$), where M is the quantum number of the J projection onto the axis Oz.

Formula (5.5.2) allows a simple modification when the EF F that acts upon the molecule is not static but oscillating $E(t) = E_0 \cos\omega t$ ($E_0 \| Oz$). In this case satellite pairs appear in the MES which are at a distance $\pm\omega$ from the position of the forbidden SL $|\Lambda_a, \nu_a, J\rangle \to |\Lambda_b, \nu_b, J \pm 2\rangle$ (Fig. 5.13). The formulas for the intensities of such satellites are obtained from (5.5.2) by the substitutions

$$F^2 \to 2^{-1}E_0^2, \qquad \Delta_b^{(J')} \to \Delta_b^{(J')} + (-1)^r\omega \quad (J' = J-1, J+2),$$

$$\Delta_a^{(J')} \to \Delta_a^{(J')} + (-1)^r\omega \quad (J' = J, J+1), \tag{5.5.6}$$

with $r = 1$ for satellites $S^{(-)}$ close to AL, while $r = 2$ for satellites $S^{(+)}$. It is assumed that the frequency ω is much less than the vibrational constants for the states a, b and does not have a resonance with the splittings $\Delta_\nu^{(J_\nu)}$ ($\nu = a, b$; $J_a = J, J+1$; $J_b = J-1, J+2$).

Using (5.5.1,2) and the formulas obtained from (5.5.2) for satellite intensities in the field $E_0 \cos\omega t$, the strengths of the static and oscillating EF in plasmas could be obtained by the experimental ratio of forbidden and allowed SL intensities. The direction of such fields could be determined by detecting the MES at different polarizer orientations and using the dependence of the forbidden SL intensities on the angle θ. Note that the proposed diagnostic method can be used not only in emission but also in fluorescence spectroscopy, depending on the manner of populating (plasma or laser) the upper level $|\Lambda_a, \nu_a, J\rangle$.

Let us evaluate typical EF strengths to be measured by the method proposed. Assume that the minimum separation between neighboring rotational levels is $\Delta/2\pi c = 1 \text{ cm}^{-1}$, and the dipole moment is $\mu = 2D$. Then supposing that the minimum detectable EF strength F_{min} satisfies the condition $\mu F_{min}/\hbar\Delta = 0.1$, we obtain $F_{min} = 3$ kV/cm. Note that the interest in EFs in molecular plasma between 1 kV/cm and several dozen kV/cm is motivated by studies of processing plasmas in high-pressure rf or microwave discharges (low-pressure discharges are characterized by relatively weak EFs, ~ 10–100 V/cm), and also by studies of breakdown of high-pressure gases in EFs. As detector molecules for measuring EFs in plasmas there could be used, for instance CO molecules (transitions $B^1\sum^+ \rightarrow A^1\Pi, C^1\sum^+ \rightarrow A^1\Pi, E^1\Pi \rightarrow A^1\Pi$) or NH molecules (transitions $c^1\Pi \rightarrow b^1\sum^+, c^1\Pi \rightarrow a^1\Delta$). Chemical reactions involving these molecules are widely used in plasma chemistry.

5.5.2 Ultra-sensitive Laser Induced Fluorescence Measurements of Weak QEFs in Low-Temperature Plasmas

Consider transitions in a diatomic polar molecule between rotational sublevels of the electronic states $^1\Pi$, $^1\sum$ (Fig. 5.14). In the absence of an external EF the transitions shown in Fig. 5.14 by solid lines are allowed, while those shown by dashed lines are forbidden. An external electric field mixes neighboring sublevels e, f of the upper rotational level, the separation Δ_{ef} between which (Λ-doubling) can be written $\Delta_{ef} = q(J+1)(J+2)$, where q is the Λ-doubling constant. Therefore when the R-branch is excited by broadband laser radiation (BLR), in the fluorescence spectrum, in addition to the R- and P-branches (with modified intensities), a forbidden line also appears, corresponding to the Q-branch. In [5.26] a new method was proposed for measuring weak constant EFs F based upon the dependence of the ratio of the Q- and P-branches I_Q/I_P on F. Relative intensities $I_Q^{(e)}$, $I_P^{(e)}$ of the Q- and P-branches were calculated as functions of the field $\boldsymbol{E}(t) = \boldsymbol{E}_0 \cos\omega t$ when the R-branch is excited by BLR [bandwidth $\Delta\omega_{LAS} \gg \max(\Delta_{ef}, \omega)$] with the polarization vector $\boldsymbol{e}_{LAS}\|\boldsymbol{E}\|Oz$.

Let a *nonresonant* field $\boldsymbol{E}(t) = \boldsymbol{E}_0 \cos\omega t$ act upon a two-level system e, f. For this case the WFs of QSs of the levels e, f were calculated using time-dependent perturbation theory. Through these WFs, we obtain the relative

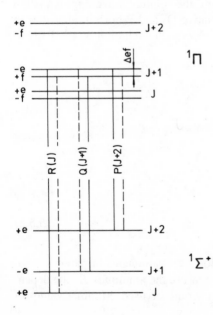

Fig. 5.14. Energy levels of a polar molecule with Λ – doubling. Allowed electric dipole transitions are indicated by *solid lines*, forbidden ones – by *dashed lines*

intensities of the Q- and P-branches in the fluorescence spectrum if the R-branch is excited by the BLR:

$$I_Q^{(e_z)} = G \sum_{M=-J}^{+J} z_{ef}^2 \rho_Q^{(z)}(M), \qquad I_P^{(e_z)} = 2 \sum_{M=-J}^{+J} \rho_P^{(z)}(M),$$

$$I_Q^{(e_x)} = G \sum_{M=-J}^{+J} z_{ef}^2 \rho_Q^{(x)}(M), \qquad I_P^{(e_x)} = \sum_{M=-J}^{+J} \rho_P^{(x)}(M),$$

$$G = E_0^2[(\Delta_{ef} - \omega)^{-2} + (\Delta_{ef} + \omega)^{-2}] \ll 1,$$

$$z_{ef}(M) = \mu M / [(J+1)(J+2)],$$

$$\rho_Q^{(z)}(M) = \frac{M^2[(J+1)^2 - M^2]}{(J+1)^2(2J+1)(2J+3)},$$

$$\rho_P^{(z)}(M) = \frac{[(J+1)^2 - M^2][(J+2)^2 - M^2]}{(2J+1)(2J+3)^2(2J+5)},$$

$$\rho_Q^{(x)}(M) = \frac{[(J+1)^2 - M^2][(J+1)(J+2) - M^2]}{2(J+1)^2(2J+1)(2J+3)},$$

$$\rho_P^{(x)}(M) = \frac{[(J+1)^2 - M]^2[(J+2)(J+3) + M^2]}{(2J+1)(2J+3)^2(2J+5)}. \tag{5.5.7}$$

Here e_x and e_z indicate the photon polarization in the fluorescence spectrum; M is the projection of J onto E_0, μ is the molecular dipole moment.

Let a *resonant* field $E(t) = E_0 \cos \omega t$ act upon a two-level system e, f: $\Delta_{ef} = \omega + \delta$, $|\delta| \ll \omega$ (δ is the detuning). For this case the WFs of the QSs of the two-level system e, f were found using the rotating wave approximation (provided $|z_{ef} E_0| \ll \omega$). As a result, the relative Q- and P-branch intensities in the fluorescence spectrum are obtained for R-branch excitation by the BLR:

$$I_Q^{(e_z)} = E_0^2 \sum_{M=-J}^{+J} \frac{z_{ef}^2}{\delta^2 + z_{ef}^2 E_0^2} \rho_Q^{(z)}(M),$$

$$I_P^{(e_z)} = \sum_{M=-J}^{+J} \frac{2\delta^2 + z_{ef}^2 E_0^2}{\delta^2 + z_{ef}^2 E_0^2} \rho_P^{(z)}(M),$$

$$I_Q^{(e_x)} = E_0^2 \sum_{M=-J}^{+J} \frac{z_{ef}^2}{\delta^2 + z_{ef}^2 E_0^2} \rho_Q^{(x)}(M),$$

$$I_P^{(e_x)} = \frac{1}{2} \sum_{M=-J}^{+J} \frac{2\delta^2 + z_{ef}^2 E_0^2}{\delta^2 + z_{ef}^2 E_0^2} \rho_P^{(x)}(M). \tag{5.5.8}$$

Expressions (5.5.7,8) can be used for local measurements of parameters of the field $E(t) = E_0 \cos \omega t$ in plasmas. To measure the amplitude E_0 the dependence of the ratio I_Q / I_P on E_0 [cf. (5.5.7)] should be used. Minimal values of

Λ-doublet splitting are $\sim 10^{-5}$–10^{-6} cm^{-1}, therefore in the range $\omega \lesssim \Delta_{ef}$ the minimal detected amplitudes could be $E_0 \sim 10$ V/cm. The direction of the vector $\boldsymbol{E}_0$ can be determined by measurements of intensities $I_Q^{(e)}$, $I_P^{(e)}$ for different photon polarizations e while the laser polarization is varied. Measurements of relatively low frequencies $\omega \sim \Delta_{ef}$ could also be carried out in the following way. By injecting a particular type of molecules into the plasma it is possible, by scanning the laser frequency over the sublevels of states $^1\Pi$ with different J, to observe at a certain J a sharp increase in the forbidden line intensity I_Q. This would correspond to the appearance of a resonance, see (5.5.8).

It should be emphasized that the proposed method can be also employed to measure substantially stronger and higher frequency QEFs, as the range of Δ_{ef} splittings is very broad in molecules (from $\sim 10^{-6}$ cm^{-1} to ~ 1 cm^{-1}). Note that the Doppler effect does not prevent measurements of QEF parameters by this method, provided that the Doppler width $\Delta\omega_{1/2}^D \ll \Delta\omega_{\text{LAS}}$.

Using these results a new method can be proposed also for measuring small doublet splittings in molecules. This can be achieved by measuring the dependence of the ratio I_Q/I_P on the frequency of the probe QEF $E(t) = E_0 \cos \omega t$. The sharp increase in I_Q/I_P at a certain ω indicates the appearance of the resonance $\omega = \Delta_{ef}$.

5.6 Frequency-Integrated Radiative Characteristics of Non-Coulomb Emitters Interacting with a Resonant Laser Field and Low-Frequency QEF

In Sect. 3.4 exactly the same problem, but for Coulomb radiators, was studied. Now we investigate a resonance transition between nondegenerate atomic levels α, β in the field of type (3.4.1) for the case in which in the nonperturbed states α, β an atom (or ion) has no average dipole moment (e.g. helium- or alkali-like emitters when $\boldsymbol{E}_{01} \parallel \boldsymbol{E}_{02}$). We shall assume that the state α is strongly perturbed by the low-frequency field $\boldsymbol{E}_2(t)$ due to coupling of the dipole with a close level α_1 ($\omega_2 \ll |\omega_{\alpha\alpha_1}^{(0)}| \ll \omega_{\alpha\beta}^{(0)}$). In addition it is assumed that the transition $\beta \to \alpha_1$ is dipole-allowed and the transition $\beta \to \alpha$ is dipole-forbidden. In this case the transition $\beta \to \alpha$ may occur with an absorption of quantum ω_1 and an absorption (or emission) of quantum ω_2. Therefore the resonance condition for the transition $\alpha \leftrightarrow \beta$ may be represented in the form

$$\omega_1 + p\omega_2 \operatorname{sign} \omega_{\alpha\alpha_1}^{(0)} = \bar{\omega}_{\alpha\beta} + \delta, \quad p = \pm 1, \tag{5.6.1}$$

where $\bar{\omega}_{\alpha\beta}$ is the separation of levels α, β with allowance for a shift of the level α in the strong field $\boldsymbol{E}_2(t)$. Because of the inequality $\omega_2 \ll |\omega_{\alpha\alpha_1}^{(0)}|$ the expressions from Sect. 5.1.2 may be used for WFs of the QSs $\Psi_\alpha(t)$ and for the shift of level α in the field $\boldsymbol{E}_2(t)$. As a result we obtain the formula for a matrix element $v_{\beta\alpha}^{(p)}$ of the transition:

$$v_{\alpha\beta}^{(1)} = \frac{(v_1)_{\beta\alpha_1}(v_2)_{\alpha_1\alpha}\exp(i\theta)}{2|(v_2)_{\alpha_1\alpha}|}\sigma_+, \qquad v_{\beta\alpha}^{(-1)} = \frac{v_{\beta\alpha}^{(1)}\sigma_-}{\sigma_+} \qquad (5.6.2)$$

with the values $\sigma_\mp$ determined by (5.1.17). In the case of the resonance (5.6.1) the expressions for the stationary difference of populations of levels α, β and for the absorbed power may be obtained from (3.4.4, 6, 7) by the substitution of a saturation parameter $G_p = 4\tau T|v_{\alpha\beta}^{(p)}|^2$ for G_r from (3.4.5).

Note that standard systems of two close levels (a, a_1) and one distant level (b), which are singled out in energy spectra of helium- or alkali-like radiators, generally consist of degenerate sublevels (with respect to magnetic quantum number m):

$$(\alpha, \alpha', \ldots) \in a, \quad (\alpha_1, \alpha_1', \ldots) \in a_1, \quad (\beta, \beta', \ldots) \in b.$$

However, in the considered conditions $(E_{01}\|E_{02})$ for each three-level subsystem $(\alpha, \alpha_1, \beta)$, $(\alpha', \alpha_1', \beta')$, ... (with fixed quantum numbers m, m', ...) the problem is solved independently and the solution is expressed by the formulas displayed above. The final result for the absorbed power is obtained by the summation of partial results for the subsystems $(\alpha, \alpha_1, \beta)$, $(\alpha', \alpha_1', \beta')$, ...

6 Non-Coulomb Emitters Under Multidimensional Dynamic EFs (Elliptically Polarized QEFs; Quasistatic EF plus QEF)

In this chapter we further discuss forbidden SL satellites. It turns out that such satellites are rather sensitive to the multidimensionality of dynamic EF (i.e. to the ellipticity degree or to a quasistatic constituent of the EF). In addition, some local features (dips, depressions) in SL profiles may be produced by joint action of static EFs and QEFs (as was shown in Sect. 4.2 for Coulomb radiators). The main results of Chap. 6 are contained in [6.1–5].

6.1 Satellites of Dipole-Forbidden Spectral Lines Caused by an Elliptically Polarized QEF

6.1.1 Three-Level Scheme, Nonresonant QEF

Consider the action of the QEF

$$E(t) = \varepsilon_0(e_x \cos \omega t + e_y \xi \sin \omega t),$$

$$\varepsilon_0 = E_0(1 + \xi^2)^{-1/2}, \quad 0 \leqslant |\xi| \leqslant 1 \tag{6.1.1}$$

upon a helium or alkali-like atom or ion, where E_0 is the amplitude and ξ the degree of ellipticity. We select from the energy spectrum of such an emitter a system of three levels i, j, k, with energies $\omega_i, \omega_j, \omega_k$, where for two neighboring levels i, j connected by a dipole matrix element, the dipole transition $j \rightarrow k$ is allowed and the transition $i \rightarrow k$ forbidden. Let the following WFs $\varphi_{nlm} \leftrightarrow i, \varphi_{n''l''m''} \leftrightarrow j, \varphi_{n'l'm'} \leftrightarrow k$ correspond to the levels i, j, k.

Consider the nonresonant influence of the field (6.1.1) on the emitter:

$$|\langle i|E(t)r|j\rangle| \ll |\omega_{ij} \pm \omega|, \qquad \omega_{pp'} \equiv \omega_p - \omega_{p'}$$

(atomic units are used: $\hbar = m_e = e = 1$). The general formula for the transition $(i, j) \rightarrow k$ is

$$I^{(e)}(\Delta\omega) = I_a^{(e)}\delta(\Delta\omega - \omega_{jk}) + S_-^{(e)}\delta(\Delta\omega - \omega_{ik} + (\text{sign } \omega_{ij})\omega)$$

$$+ S_+^{(e)}\delta(\Delta\omega - \omega_{ik} - (\text{sign } \omega_{ij})\omega), \tag{6.1.2}$$

where $I_a^{(e)}$ is the allowed line intensity, $S_-^{(e)}$, $S_+^{(e)}$ are the intensities of near and far forbidden line satellites, and e is the unit polarization vector of the emitted photons. We present the results of the calculation of the component intensities in the spectrum (6.1.2) (in the Coulomb approximation) for the following situations allowed by the selection rules:

1) $l'' = l - 1, l' = l - 2$

$$I_a^{(e_x)} = I_a^{(e_y)} = I_a^{(e_z)} = 3^{-1}(l-1)\left|R_{n'',l-1}^{n',l-2}\right|^2,$$

$$S_\pm^{(e_x)} = \frac{(4+3\xi^2)E_0^2}{120(1+\xi^2)(|\omega_{ij}|\pm\omega)^2} \cdot \frac{l(l-1)}{2l-1}\left|R_{n'',l-1}^{n',l-2}\right|^2\left|R_{n,l}^{n'',l-1}\right|^2,$$

$$S_\pm^{(e_y)} = \frac{3+4\xi^2}{4+3\xi^2}S_\pm^{(e_x)}, \qquad S_\pm^{(e_z)} = \frac{3(1+\xi^2)}{4+3\xi^2}S_\pm^{(e_x)}. \qquad (6.1.3)$$

2) $l'' = l + 1, l' = l + 2$

$$I_a^{(e_x)} = I_a^{(e_y)} = I_a^{(e_z)} = 3^{-1}(l+2)\left|R_{n'',l+1}^{n',l+2}\right|^2,$$

$$S_\pm^{(e_x)} = \frac{(4+3\xi^2)E_0^2}{120(1+\xi^2)(|\omega_{ij}|\pm\omega)^2} \cdot \frac{(l+1)(l+2)}{2l+3}\left|R_{n,l}^{n'',l+1}\right|^2\left|R_{n'',l+1}^{n',l+2}\right|^2,$$

$$S_\pm^{(e_y)} = \frac{3+4\xi^2}{4+3\xi^2}S_\pm^{(e_x)}, \qquad S_\pm^{(e_z)} = \frac{3(1+\xi^2)}{4+3\xi^2}S_\pm^{(e_x)}. \qquad (6.1.4)$$

3) $l'' = l + 1, l' = l$

$$I_a^{(e_x)} = I_a^{(e_y)} = I_a^{(e_z)} = 3^{-1}(l+1)\left|R_{n'',l+1}^{n',l}\right|^2,$$

$$S_\pm^{(e_x)} = \frac{[8l^2+16l+10+l(6l+7)\xi^2](l+1)E_0^2}{120(1+\xi^2)(|\omega_{ij}|\pm\omega)^2(2l+1)(2l+3)}\left|R_{n,l}^{n'',l+1}\right|^2\left|R_{n'',l+1}^{n',l}\right|^2,$$

$$S_\pm^{(e_y)} = \frac{l(6l+7)+(8l^2+16l+10)\xi^2}{8l^2+16l+10+l(6l+7)\xi^2}S_\pm^{(e_x)},$$

$$S_\pm^{(e_z)} = \frac{(1+\xi^2)l(6l+7)}{8l^2+16l+10+l(6l+7)\xi^2}S_\pm^{(e_x)}. \qquad (6.1.5)$$

4) $l'' = l - 1, l' = l$

$$I_a^{(e_x)} = I_a^{(e_y)} = I_a^{(e_z)} = 3^{-1}l\left|R_{n'',l-1}^{n',l}\right|^2,$$

$$S_\pm^{(e_x)} = \frac{[8l^2+2+(l+1)(6l-1)\xi^2]lE_0^2}{120(1+\xi^2)(|\omega_{ij}|\pm\omega)^2(2l+1)(2l-1)}\left|R_{n,l}^{n'',l-1}\right|^2\left|R_{n'',l-1}^{n',l}\right|^2,$$

$$S_\pm^{(e_y)} = \frac{(l+1)(6l-1)+(8l^2+2)\xi^2}{8l^2+2+(l+1)(6l-1)\xi^2}S_\pm^{(e_x)},$$

$$S_\pm^{(e_z)} = \frac{(1+\xi^2)(l^2-1)}{8l^2+2+(l+1)(6l-1)\xi^2}S_\pm^{(e_x)}. \qquad (6.1.6)$$

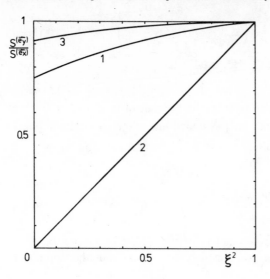

Fig. 6.1. The ratio of emission intensities of a single satellite of a helium-like ion corresponding to e_y and e_x polarizations vs. the ellipticity degree ξ of the QEF: Curve 1: cases 1 and 2 (6.1.3,4) for arbitrary l; curve 2: case 3 (6.1.5) for $l = 0$; curve 3: case 4 (6.1.6) for $l = 3$

In expressions (6.1.3–6) $R^{n',l'}_{n'',l''}$ is an integral over the radial WFs:

$$R^{n'l'}_{n''l''} = \int\limits_0^\infty dr\, r^3 R_{n'',l''}(r) R_{n',l'}(r).$$

Note that for $\xi \to 0$ these results coincide with the results of [6.6]. Using (6.1.3–6) one can experimentally determine the ellipticity degree ξ of the field (6.1.1) by the ratio of intensities of a single satellite, detected in two polarizations (Fig. 6.1), and E_0 can be determined by the ratio of satellite intensities to the intensity of an allowed line (which remains unpolarized).

6.1.2 Three-Level Scheme, Resonant QEF

Consider the action of the QEF (6.1.1) upon an emitter in the resonance case $\omega_{ji} = \omega - \sigma, |\sigma| \ll \omega$, where for definiteness we set $\omega_{ji} > 0$, and σ is the detuning. The radiation spectrum for the transition $(nS, n''P) \to n'S$ has the form

$$I^{(e_x)}(\Delta\omega) \propto (1+\xi^2)^{-1}\left\{\xi^2\delta(\Delta\omega - \omega_{jk}) + 2^{-1}\right.$$
$$\times\left[\left(1 + \frac{\sigma}{2\Omega_1}\right)\delta\left(\Delta\omega - \omega_{jk} + \Omega_1 - \frac{\sigma}{2}\right)\right.$$
$$\left.\left. + \left(1 - \frac{\sigma}{2\Omega_1}\right)\delta\left(\Delta\omega - \omega_{jk} - \Omega_1 - \frac{\sigma}{2}\right)\right]\right\},$$

$$I^{(e_y)}(\Delta\omega) \propto (1+\xi^2)^{-1}\left\{\delta(\Delta\omega - \omega_{jk}) + 2^{-1}\xi^2\right.$$
$$\times\left[\left(1 + \frac{\sigma}{2\Omega_1}\right)\delta\left(\Delta\omega - \omega_{jk} + \Omega_1 - \frac{\sigma}{2}\right)\right.$$

$$+ \left(1 - \frac{\sigma}{2\Omega_1}\right) \delta \left(\Delta\omega - \omega_{jk} - \Omega_1 - \frac{\sigma}{2}\right)\Bigg]\Bigg\}, \tag{6.1.7}$$

$$I^{(e_z)}(\Delta\omega) \propto \delta(\Delta\omega - \omega_{jk}), \qquad \Omega_1 \equiv \Omega_1^{(S,P)},$$

where the Rabi frequency is $\Omega_1^{(S,P)} = [\sigma^2/4 + E_0^2 (R_{n,0}^{n'',1})^2/12]^{1/2}$. In the case of resonance interactions of the QEF (6.1.1) with the pair of levels nP, $n''D$, three different Rabi frequencies appear, $\Omega_{1,2,3}^{(P,D)}$, which indicate a spectral component splitting, where $\Omega_1^{(P,D)}$ is obtained from $\Omega_1^{(S,P)}$ by the substitution $R_{n,0}^{n'',1} \rightarrow (3/5)^{1/2} R_{n,1}^{n'',2}$ and

$$\Omega_{2,3}^{(P,D)} = \left[\frac{\sigma^2}{4} + \frac{7 \pm [1 + 96(\xi + \xi^{-1})^{-2}]^{1/2}}{120} E_0^2 (R_{n,1}^{n'',2})^2\right]^{1/2}. \tag{6.1.8}$$

Thus in the resonance case the parameters of the QEF (6.1.1) may be determined either from the spectral component splitting or from the component intensities (6.1.7,8).

6.1.3 Four-Level Scheme

Consider the influence of a QEF (6.1.1) on the emission spectrum $(2P_{1/2}, 2S_{1/2}, 2P_{3/2}) \rightarrow 1S_{1/2}$ of a hydrogen-like ion with a nuclear charge Z. Designate: $0 \leftrightarrow 1S_{1/2}$, $1 \leftrightarrow 2S_{1/2}$, $2 \leftrightarrow 2P_{3/2}$, $3 \leftrightarrow 2P_{1/2}$.

In the nonresonant case $(|\omega \pm \omega_{k1}| > |r_{1k}E(t)|, k = 2, 3)$ the spectrum has the form

$$I_a(\Delta\omega) = 2\delta(\Delta\omega - \omega_{20}) + \delta(\Delta\omega - \omega_{30}) + S_-^{(e)}\delta(\Delta\omega - \omega_{10} - \omega)$$

$$+ S_+^{(e)}\delta(\Delta\omega - \omega_{10} + \omega),$$

$$S_{\mp}^{(e_x)} = \frac{3E_0^2}{4Z^2(1+\xi^2)}$$

$$\times \left(\frac{4+\xi^2}{(\omega_{21} \mp \omega)^2} + \frac{1+\xi^2}{(\omega_{13} \mp \omega)^2} - \frac{4-2\xi^2}{(\omega_{21} \mp \omega)(\omega_{13} \pm \omega)}\right),$$

$$S_{\mp}^{(e_y)} = \frac{3E_0^2}{4Z^2(1+\xi^2)}$$

$$\times \left(\frac{4\xi^2+1}{(\omega_{21} \mp \omega)^2} + \frac{1+\xi^2}{(\omega_{13} \pm \omega)^2} - \frac{4\xi^2-2}{(\omega_{21} \mp \omega)(\omega_{13} \pm \omega)}\right),$$

$$S_{\mp}^{(e_z)} = 3E_0^2(2Z)^{-2}[(\omega_{21} \mp \omega)^{-1} + (\omega_{13} \pm \omega)^{-1}]^2. \tag{6.1.9}$$

In the resonant case $\omega_{21} = \omega - \sigma$ $(|\sigma| \ll \omega)$ the spectrum $(2S_{1/2}, 2P_{3/2}) \rightarrow 1S_{1/2}$ can be expressed as

$$I^{(e)}(\Delta\omega) = A^{(e)}\delta(\Delta\omega - \omega_{20}) + \sum_{s,p=1}^{2} B_p^{(e)}[2\tilde{\Omega}_p^2 + (-1)^s \tilde{\Omega}_p \sigma]^{-1}$$

$$\times \delta(\Delta\omega - \omega_{20} + \sigma/2 + (-1)^s \tilde{\Omega}_p);$$

$$A^{(e_x)} = \xi^2 A^{(e_y)}, \qquad A^{(e_y)} = 3^{-1}(1 + \xi^2)[(1 + \xi^2)^2 - \xi^2]^{-1},$$

$$A^{(e_z)} = (1 + \xi^4)(1 + \xi^2)^{-1} A^{(e_y)}, \qquad B_p^{(e_x)} = [2 - (-1)^p \xi]^2 V,$$

$$B_p^{(e_y)} = [2\xi - (-1)^p]^2 V, \qquad B_p^{(e_z)} = [1 + (-1)^p \xi]^2 V,$$

$$\tilde{\Omega}_p = \{\sigma^2/4 + (3/2)Z^{-2}(1 + \xi^2)^{-1} E_0^2 [1 + \xi^2 - (-1)^p \xi]\}^{1/2},$$

$$V = (E_0^2/12)Z^{-2}(1 + \xi^2)^{-1}. \tag{6.1.10}$$

Using (6.1.9,10) one can measure ξ and E_0 in high-temperature plasmas as in the procedure described in Sects. 6.1.2,3.

6.2 Joint Action of QEF and Intraplasmic Quasistatic EF on Non-Coulomb Emitters

6.2.1 Strong Influence of a Quasistatic EF on Satellites of Dipole-Forbidden Lines

Up to now, the possibility of a strong influence of a moderate static EF F on the spectrum has been largely ignored. It was assumed that their only effect was the emergence of a forbidden line with an intensity proportional to F^2.

However, in atoms for which the Stark effect is linear (e.g., hydrogen), under joint action of the F and $E_0 \cos \omega t$ fields, a drastic change from the spectrum seen in a $E_0 \cos \omega t$ field only is observed. It is clear from the outset that it is necessary to take F into account when it is very strong, so that a transition to the linear Stark effect occurs. Here we demonstrate the striking fact that it is important to take F into account in forbidden SL satellites even at comparatively low F when the Stark effect is still quadratic.

Consider in the energy spectrum of a helium- or alkali-like emitter a three-level system with the energies $\omega_0^{(0)}$, $\omega_1^{(0)}$, $\omega_2^{(0)}$ and the corresponding WFs $\varphi_{i,m}$ ($i = 0, 1, 2$; m is the magnetic quantum number), while $|\omega_{21}^{(0)}| \equiv |\omega_2^{(0)} - \omega_1^{(0)}| \ll |\omega_1^{(0)} - \omega_0^{(0)}|$. Each of these levels is m-degenerate. Suppose dipole transitions $1 \rightarrow 0, 1 \rightarrow 2$ are allowed, and $2 \rightarrow 0$ is forbidden (which corresponds to $l_1 = l; l_0 = l - 1; l_2 = l + 1; l$ is the angular momentum). Assume that in a coordinate system $Oxyz$ the vectors F and E_0 are $F = (0, 0, F)$, $E_0 = (E_{0x}, 0, E_{0z})$.

We solve the Schrödinger equation

$$i \partial\psi/\partial t = [H_0 + zF + (zE_{0z} + xE_{0x}) \cos \omega t]\psi, \tag{6.2.1}$$

where H_0 is the unperturbed Hamiltonian $H_0 \varphi_i^{(m)} = \omega_i^{(0)} \varphi_i^{(m)}$ (atomic units $\hbar = e = m_e = 1$ are used), employing the basis of eigen-WFs of the Hamiltonian $H_0 + zF$:

$$\chi_1^{(m)} = \varphi_1^{(m)} \cos(\beta_m/2) - \varphi_2^{(m)} \sin(\beta_m/2),$$

$$\chi_2^{(m)} = \varphi_2^{(m)} \cos(\beta_m/2) + \varphi_1^{(m)} \sin(\beta_m/2). \tag{6.2.2}$$

Here $\tan \beta_m = 2\langle \varphi_1^{(m)} | z | \varphi_2^{(m)} \rangle F / \omega_{21}^{(0)}$, and the new energy levels are

$$\omega_{21}^{(m)} = 2^{-1}[\omega_1^{(0)} + \omega_2^{(0)} \pm \omega_{21}^{(m)}(F)],$$

$$\omega_{21}^{(m)}(F) = \omega_{21}^{(0)}(1 + \tan^2 \beta_m)^{1/2}. \tag{6.2.3}$$

Solving equation (6.2.1) we obtain the following QSs

$$\psi_1^{(m)}(t) = \left[\chi_1^{(m)} - \sum_{k=\pm 1} \Phi_{1,k} \exp(-ik\omega t) \right] \exp(-i\omega_{10}^{(m)}t),$$

$$\Psi_2^{(m)}(t) = \left[\chi_2^{(m)} - \sum_{k=\pm 1} \Phi_{2,k} \exp(-ik\omega t) \right] \exp(-i\omega_{20}^{(m)}t), \tag{6.2.4}$$

where $\Phi_i^{(m)}$ are some linear combinations of the functions $\chi_j^{(m')}$ obtained from the perturbation theory. The spontaneous emission spectrum with the polarization vector parallel to E_0 for the transition $(n_1, l; n_2, l+1) \to (n_0, l-1)$ is

$$I_\parallel(\Delta\omega, F, \alpha) = \sum_m \left\{ I_a^{(m)} \delta(\Delta\omega - \omega_1^{(m)}) + I_f^{(m)} \delta(\Delta\omega - \omega_2^{(m)}) \right.$$

$$\left. + \sum_{k=\pm 1} \left[S_{a,k}^{(m)} \delta(\Delta\omega - \omega_1^{(m)} + k\omega) + S_{f,k}^{(m)} \delta(\Delta\omega - \omega_2^{(m)} - k\omega) \right] \right\}, \tag{6.2.5}$$

where α is the angle between the vectors F and E_0. Component intensities in (6.2.5) are functions of F and α. We give them only for the particular case of $\alpha = 0$:

$$I_a \approx \cos^2(\beta/2) \approx 1 - I_f,$$

$$S_{f,\mp 1} = \frac{z_{12}^2 E_0^2}{4} \left(\frac{(\sin\beta)\sin(\beta/2)}{\omega} \mp \frac{(\cos\beta)\cos(\beta/2)}{|\omega_{21}(F)| \mp \omega} \right)^2,$$

$$S_{a,\mp 1} = \frac{z_{12}^2 E_0^2}{4} \left(\frac{(\sin\beta)\cos(\beta/2)}{\omega} \pm \frac{(\cos\beta)\sin(\beta/2)}{|\omega_{21}(F)| \mp \omega} \right)^2. \tag{6.2.6}$$

In (6.2.6) I_a is the allowed line intensity, I_f is the forbidden line intensity; $S_{f,-1}$, $S_{f,+1}$ are the intensities of the near and far forbidden line satellites. In addition, if $F \neq 0$, two allowed line satellites located at a distance $\pm\omega$ from the allowed line position are present in the spectrum with the intensities $S_{a,-1}$ and $S_{a,+1}$ proportional to E_0^2. The appearance of these satellites is a striking new feature. Indeed, allowed line satellites known at $F = 0$ have an intensity proportional to E_0^4 and are placed at $\pm 2\omega$ from the allowed line position. It follows from (6.2.6) that if F is close to zero, the main contribution to the intensity of the forbidden line satellites comes from the terms containing $[\omega_{21}(F) \mp \omega]^{-1}$ (Baranger–Mozer type terms). However, with the increase of F the contribution of these terms decreases, but that of the terms containing $1/\omega$

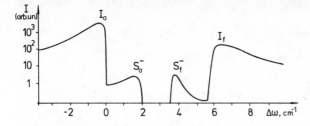

Fig. 6.2. Helium spectral lines corresponding to the transitions $2^1P - (4^1D, 4^1F)$ under the joint action of a QEF $E_0 \cos \omega t$ and Holtsmark microfield ($E_0 = 1$ kV/cm, $N_i = 10^{14}$ cm^{-3}): I_a is the allowed line, I_f the forbidden line, S_a^- and S_f^- are near satellites of allowed and forbidden lines, respectively

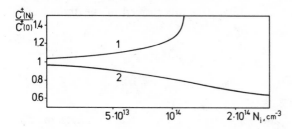

Fig. 6.3. A Ratio $C^{\pm}(N_i) = \langle S_f^{\pm} \rangle / \langle I_a \rangle$ of satellite and allowed line intensities, calculated assuming the presence of Holtsmark microfield, divided by the same ratio $C^{\pm}(0) = S_f^{\pm}/I_a$, calculated without Holtsmark microfield. The results are shown for the same transitions as in Fig. 6.2 at $E_0 = 1$ kV/cm

increases. At sufficiently high F, intensities of forbidden line satellites tend to equalize: $S_{f,-1} \approx S_{f,+1} \approx E_0^2 z_{12}^2/(8\omega^2)$; in this case the forbidden line satellites are analogous to the Blochinzew satellites in the hydrogen atom spectrum in the field $E_0 \cos \omega t$ (Sect. 3.1).

To obtain the resulting SL profile for a general case (for arbitrary α), one needs to average over the angle α and the Holtsmark distribution function $W_H(F/F_H)$ [6.7] of the field F

$$\langle I_{\parallel}(\Delta \omega) \rangle = \int_0^{\infty} d(F/F_H) W_H(F/F_H) \int_{-1}^{+1} d(\cos \alpha) 2^{-1} I_{\parallel}(\Delta \omega, F, \alpha). \quad (6.2.7)$$

Figure 6.2 shows the calculated profile $\langle I_{\parallel}(\Delta \omega) \rangle$ of the helium SL corresponding to the transitions $2^1P - (4^1D, 4^1F)$. It is seen that the unusual satellite separated by the value of ω from the allowed SL is just as intense as the usual forbidden SL satellite.

It is important to emphasize that even relatively small Holtsmark microfields (with a normal field $F_H = 2.60 e N_i^{2/3} \lesssim E_0/3$) substantially modify the intensities of usual forbidden SL satellites and consequently the resulting E_0 values found by the Baranger–Mozer method (Sect. 5.1.1). This surprising result is illustrated in Fig. 6.3 which clearly shows dramatic changes in the ratio of satellite and allowed line intensities caused by Holtsmark microfields (especially for the far satellite S_f^+). This implies that, for the most part, the results obtained by observations of Baranger–Mozer type satellites should be reinterpreted.

6.2.2 Intra-Stark Spectroscopy[1]

Helium- or alkali-like emitter. We select from the energy spectrum of a helium- or alkali-like emitter a "standard" system of three levels 0, 1, 2, as described in Sect. 5.1.1 (Fig. 5.1). We consider the case in which the unperturbed distance Δ between the levels 1 and 2 is smaller than the frequency ω of the field $E_0 \cos \omega t$. In this case for some group of radiators a quasistatic EF F, moving the levels 1 and 2 apart, may tune them to be in resonance with the field $E_0 \cos \omega t : \bar{\Delta}(F) = \omega$.[2] This leads to an additional splitting of the levels of $\delta\omega \sim (\pm z_{12} E_0/2)$, where z_{12} is the dipole matrix element between the levels 1 and 2. As a result, in the averaged (over quasistatic EF) profiles of an allowed SL, a forbidden SL and its far satellite, dips are formed at the frequencies ω_a^{dip}, ω_f^{dip} and $\omega_{S_+}^{dip}$, respectively, where

$$\omega_a^{dip} = \omega_0 + (\text{sign}\,\omega_{21}^{(0)})\omega/2, \qquad \omega_f^{dip} = \omega_0 - (\text{sign}\,\omega_{21}^{(0)})\omega/2,$$

$$\omega_{S_+}^{dip} = \omega_0 - 3(\text{sign}\,\omega_{21}^{(0)})\omega/2, \qquad \omega_0 \equiv (\omega_{10}^{(0)} + \omega_{20}^{(0)})/2; \qquad (6.2.8)$$

the width of the dips is $\Omega \sim z_{12} E_0$.

We analyze in detail the situation where the QEF frequency ω coincides with the electron plasma frequency $\omega_{pe} = (4\pi e^2 N_e/m_e)^{1/2}$ and the static EF F is characterized by the known [6.7] Holtsmark distribution $W_H(F, F_H)$, $F_H = 2.60 e N_e^{2/3}$. Then the resonant value of F [satisfying the condition $\bar{\Delta}(F) = \omega$] may be written

$$F = K(N_e) e N_e^{2/3}, \qquad K(N_e) = \left(\frac{4\pi e^2}{m_e} N_e - \Delta^2\right)^{1/2} \left(\frac{2\tilde{z}_{12}\hbar}{m_e} N_e^{2/3}\right)^{-1},$$

$$(6.2.9)$$

where $\tilde{z}_{12} \equiv z_{12}/a_0$; a_0 is the Bohr radius. It is easy to show that at $m_e \Delta^2/(4\pi e^2) \leqslant N_e < \tilde{N}_e$ the function $K(N_e)$ increases monotonically from 0 to K_{max} and at $N_e > \tilde{N}_e$ decreases monotonically from K_{max} to 0, the values of $\tilde{N}_e$, K_{max} being $\tilde{N}_e = (\hbar\Delta/I_H)^2/4\pi a_0^3$, $K_{max} = 3^{1/2}(\pi/2)^{2/3} \times (I_H/\hbar\Delta)^{1/3}/\tilde{z}_{12}$ (I_H is the hydrogen ionization potential).

For the dips at the frequencies given in (6.2.8) to be visible it is necessary that the value $F = K_e N_e^{2/3}$ not differ significantly from the most probable field value $F_m \approx 4.21 e N_e^{2/3}$. In particular, we demand that the condition $2 \lesssim K \lesssim 8$ [under which $W_H(F, F_H) \geqslant W_H(F_m, F_H)/2$] be fulfilled. Then from (6.2.9) it is easy to determine the range of densities N_e where these dips can be observed and used for diagnostics. For instance, in the emission spectrum near

---

[1] The main results of this section have been obtained jointly with V.P. Gavrilenko.

[2] A high concentration of radiators for which two sublevels (or more than two as in hydrogen) are tuned to be in resonance with the Langmuir wave, at first glance, might appear to lead to strong damping of such a wave. But the estimations show that in practice this is improbable (Appendix E).

the SL of He I at 492.2 nm, for the radiative transitions $2^1P - (4^1D, 4^1F)$, the dips may be conveniently utilized for plasma diagnostics in the range $N_e \approx (0.7-2) \times 10^{16}$ cm^{-3}. At greater densities, $N_e \approx (2-10) \times 10^{16}$ cm^{-3}, dips may also be used for diagnostics, but with the additional inclusion of the influence of the 4^1P level on the two-level subsystem 4^1D, 4^1F.

Hydrogen-like ions. Consider the evolution of a two-level system n $S_{1/2}$, n $P_{1/2}$ of a hydrogen-like ion (with nuclear charge Z) in the EF $E(t) = F + E_0 \cos \omega t$. Let the single-quantum resonance condition be fulfilled:

$$\bar{\Delta}(F) = \omega + \delta, \quad |\delta| \ll \omega;$$
$$\bar{\Delta}(F) = \Delta(1 + \tan^2 \beta^\pm)^{1/2}, \quad \tan \beta^\pm = 2z_{12}^\pm F/\Delta,$$
$$z_{12}^\pm = \pm n(n^2 - 1)^{1/2}/(2Z). \tag{6.2.10}$$

Then, after solving the Schrödinger equation in the resonance approximation, we finally obtain the following expression for the spontaneous emission spectrum of the transition $n'S_{1/2} - (nS_{1/2}, nP_{1/2})$:

$$I(\Delta\omega) = \sum_{k=1}^{2}[C_k\delta(\Delta\omega + \omega/2 + (-1)^{k+1}(\delta^2 + \Omega^2)^{1/2}/2)$$
$$+ S_k\delta(\Delta\omega - \omega/2 + (-1)^{k+1}(\delta^2 + \Omega^2)^{1/2}/2)],$$
$$S_k = \Omega^2 \sin^2(\beta/2)\{\Omega^2 + [\delta + (-1)^{k+1}(\delta^2 + \Omega^2)^{1/2}]^2\}^{-1},$$
$$C_k = \cos^2(\beta/2) - S_k\cot^2(\beta/2), \tag{6.2.11}$$

where $\Omega = |z_{12}|(E_{0\perp}^2 + E_{0\parallel}^2 \cos^2 \beta)^{1/2}$ is the Rabi frequency ($\Omega \ll \omega$), $E_{0\parallel}$ and $E_{0\perp}$ are the components of E_0 parallel and transverse to F; $\Delta\omega$ is counted from the halfsum of frequencies of transitions $n'S_{1/2} - nS_{1/2}, n'S_{1/2} - nP_{1/2}$. In (6.2.11) (which corresponds to the emission into the solid angle 4π) the terms with C_k describe the splitting of the allowed SL $n'S_{1/2} - nP_{1/2}$, the terms with S_k give the splitting of the forbidden (at $F = 0$) SL $n'S_{1/2} - nS_{1/2}$. From (6.2.11) it follows that at any F there is no emission in the interval $|\Delta\omega + \omega/2| < v/2$, $|\Delta\omega - \omega/2| < v/2$, where $v \equiv \min_{\delta}(\delta^2 + \Omega^2(\delta)) \approx \Omega(F_{res})$ [F_{res} is the "resonant" value of a static EF: $\bar{\Delta}(F_{res}) = \omega$]. Therefore in these intervals dips arise in the resulting profiles of the allowed ($n'S_{1/2} - nP_{1/2}$) and forbidden ($n'S_{1/2} - nS_{1/2}$) lines [after averaging of $I(\Delta\omega)$ from (6.2.11) over a distribution of quasistatic fields F]. The width of dips is estimated to be

$$\Omega^{dip} \sim (\langle\Omega^2(F_{res})\rangle)^{1/2} = |z_{12}|E_0\{[(\Delta/\omega)^2 + 2]/3\}^{1/3}. \tag{6.2.12}$$

Although in this treatment we neglected the shift of the levels $nS_{1/2}, nP_{1/2}$ (in the field F) caused by other levels (which is valid at relatively small F), dips, due to the resonance $\bar{\Delta}(F) = \omega$, will appear even with an allowance for such a shift – only the form of the function $\bar{\Delta}(F)$ will change. Note that in the case of the L_α spectrum the influence of the $2 P_{3/2}$ level on the two-level system $2S_{1/2}, 2P_{1/2}$ is small when $F \lesssim \tilde{F} = 10^3$ V/cm.

At greater F (where the fine structure and the Stark splitting in the field F are comparable) when $n = 2$, in addition to dips on L_α profiles caused by a resonance in the two-level subsystem $2S_{1/2}$, $2P_{1/2}$, other dips may also be visible due to resonances with sublevels of the level $2P_{3/2}$ (including dips on the profile of the $1S_{1/2} - 2P_{3/2}$ line). For example, for an Na XI ion under the action of a Nd-laser ($\omega = \omega_{Nd} \approx 1.78 \times 10^{15}$ s^{-1}) the following resonances with the field $E_0 \cos \omega t$ are possible: at $F = 3.2 \times 10^8$ V/cm between the levels $2P_{1/2}$ and $2P_{3/2}(m_j = \pm 1/2)$, at $F = 4.7 \times 10^8$ V/cm between the levels $2P_{1/2}$ and $2P_{3/2}(m_j = \pm 3/2)$, at $F = 7.6 \times 10^8$ V/cm between the levels $2S_{1/2}$ and $2P_{3/2}(m_j = \pm 1/2)$, at $F = 7.8 \times 10^8$ V/cm between the levels $2P_{1/2}$ and $2S_{1/2}$.

Experimental detection of dips will allow the determination of the amplitude of a QEF E_0 (from the dip halfwidth $\Omega^{dip} \propto E_0$) and in the case of Langmuir oscillations the concentration N_e can also be found, from the dip separation ω_{pe}.

6.2.3 Joint Action of a Quasistatic EF and a High-Frequency QEF on a Hydrogen-like Ion. Fine Structure and Lamb Shift. Local Measurements of Amplitude Angular Distributions of Low-Frequency Plasma Turbulence

Let us consider the transformation of the L_α spectrum of a hydrogen-like ion in the field $E(t) = E_0 \cos \omega t + F$. Let the vectors F and E_0 in the $Oxyz$ system be $E_0 = (0, 0, E_0)$, $F = (F_x, 0, F_z)$. We denote the nonperturbed states $|nljm\rangle$ of levels $2P_{3/2}$, $2S_{1/2}$, $2P_{1/2}$ by $|\pm 1\rangle = |201/2 \pm 1/2\rangle$, $|\pm 2\rangle = |213/2 \pm 1/2\rangle$, $|\pm 3\rangle = |211/2 \pm 1/2\rangle$, $|\pm 4\rangle = |213/2 \pm 3/2\rangle$. Let the energy $\omega_k (k = 1, 2, 3, 4)$ correspond to the state $|\pm k\rangle$. Quasienergies μ_k and WFs of QSs were found analytically in Sect. 5.3.1. Recall that the condition of validity of these results is that the field $E_0 \cos \omega t$ be an intense or high-frequency field with respect to the sublevels with $n = 2$:

$$\max \left((\xi E_0/\omega)^{1/2}, \omega \right) \gg \left(\omega_{21} z_{12}^2 + \omega_{31} z_{13}^2 \right) / \xi^2,$$

$$\xi \equiv \left(z_{12}^2 + z_{13}^2 \right)^{1/2}, \tag{6.2.13}$$

where $\omega_{kk'} \equiv \omega_k - \omega_{k'}$, and $z_{kk'} \equiv \langle k|z|k'\rangle$ are dipole moment matrix elements (here and below $\hbar = m_e = e = 1$). Figure 6.4 shows, using the example of the O VIII ion in an Nd laser field, the typical dependence of quasienergies for the $n = 2$ states on the amplitude E_0.

Consider perturbations of QSs, formed in the field $E_0 \cos \omega t$, under the action of the field F. In this case, as seen in Fig. 6.4, two qualitatively different situations are possible: (a) all quasienergies μ_k are well separated from each other, (b) the quasienergies are degenerate, $\mu_1 \approx \mu_2$.

The results obtained are the following. In case (a) the QSs of the level n = 2 in the field $E(t)\tilde{\mu}_k = \mu_k + \Delta\mu_k$ include the term $\Delta\mu_k$, which is quadratic in the projections of the field F:

$$\Delta\mu_1 = -(\Delta\mu_2 + \Delta\mu_3 + \Delta\mu_4),$$

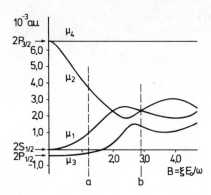

Fig. 6.4. Quasienergies μ_k (for $n = 2$) vs. QEF amplitude E_0 for the O VIII ion in an Nd laser field $E_0 \cos \omega t$

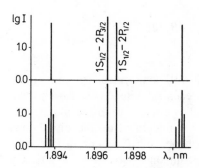

Fig. 6.5. A section of the O VIII ion L_α spectrum. Upper part: in a Nd laser field ($\omega = 1.78 \times 10^{15} \ s^{-1}$) with $E_0 = 6 \times 10^8$ V/cm (main components and nearest satellites are shown). Lower part: in a field $E_0 \cos \omega t + F$, $F \perp E_0$, $E_0 = 6 \times 10^8$ V/cm, $F = 3 \times 10^7$ V/cm

$$\Delta\mu_2 = z_{13}^2[2g^2 F_z^2 + f^2 J_0^2(B) F_x^2]/[2(\mu_2 - \mu_1)],$$
$$\Delta\mu_3 = z_{13}^2[2f^2 F_z^2 + g^2 J_0^2(B) F_x^2]/[2(\mu_3 - \mu_1)],$$
$$\Delta\mu_4 = 3z_{13}^2 J_0^2(B) F_x^2/[2(\mu_4 - \mu_1)];$$
$$f \equiv (1 + 2^{1/2}A)(1 + A^2)^{-1/2},$$
$$g \equiv (2^{1/2} - A)(1 + A^2)^{-1/2}, \quad B \equiv \xi E_0/\omega, \tag{6.2.14}$$

where $J_k(x)$ are Bessel functions; the quantity A has been defined in Sect. 5.3.1. Figure 6.5 shows the changes in the L_α spectrum (with z-polarization) for the O VIII ion, typical for case (a).

In case (b) an additional component splitting appears in the L_α spectrum which is linear in the projections of the field F. Neglecting corrections which are quadratic in F, we represent the L_α spectrum (with Z-polarisation) in the form

$$I_z(\Delta\omega) = \sum_{i=\pm1, \pm3} \sum_{p=-\infty}^{+\infty} I_i^{(p)} \delta(\Delta\omega - p\omega - \tilde{\mu}_i), \ \tilde{\mu}_i = \mu_i + \Delta\mu_i,$$

$$\Delta\mu_{\pm3} = 0, \ \Delta\mu_{\pm1} = \pm z_{13}[F_z g + 2^{-1/2} F_x J_0(B) f]. \tag{6.2.15}$$

In (6.2.15) and below the frequency $\Delta\omega$ is relative to the nonperturbed frequency of the transition $2S_{1/2} \to 1S_{1/2}$. The intensities of $I_i^{(p)}$ components

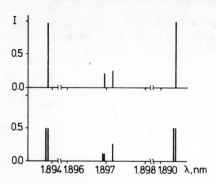

Fig. 6.6. The same as in Fig. 6.5 but for a degenerate case. Upper part: in a field $E_0 \cos \omega t$, $E_0 = 8.9 \times 10^8$ V/cm. Lower part: in a field $E_0 \cos \omega t + F$, $E_0 \perp F$, $E_0 = 8.9 \times 10^8$ V/cm, $F = 10^7$ V/cm

in (6.2.15) are

$$I_{\pm 1}^{(2p)} = g^2 J_{2p}^2(B)/2, \qquad I_{\pm 1}^{(2p+1)} = 3 J_{2p+1}^2(B)/2,$$
$$I_{\pm 3}^{(2p)} = f^2 J_{2p}(B)/2, \qquad I_{\pm 3}^{(2p+1)} = 0. \qquad (6.2.16)$$

Figure 6.6 shows the L_α spectrum of the O VIII ion, typical for case (b). It should be emphasized that the field F splits each satellite $\Delta \omega = \pm \omega$ into two components. To obtain the resulting L_α profiles emitted from the plasma, L_α spectra obtained in cases (a) and (b) must be integrated over a distribution of quasistatic plasma fields F. Parameters of the field $E_0 \cos \omega t + F$ can be measured either from the shifts of spectral components or by their intensity ratios.

It should be noted that if the field $E_0 \cos \omega t$ is the laser (or maser) radiation field and it is possible to vary its parameters (amplitude, frequency, polarization) for diagnostic purposes, then the amplitude-angular distribution of quasistatic fields F in plasmas can be easily measured. In such measurements the quasistatic broadening of odd satellites $\Delta \omega \approx (2p+1)\omega$ in the degenerate case (b) should be used, for the following reasons: 1) in case (b) the broadening occurs in a regime corresponding to a linear Stark effect (with respect to F); 2) odd satellites in the field F are split completely; 3) the main components $\Delta \omega = 0$ are not superimposed upon the satellites $\Delta \omega \approx (2p+1)\omega$.

Note in conclusion that effects analogous to those considered above also exist in the spectra of helium- and alkali-like emitters, as for example for the transitions (2S, 2P) $\rightarrow$ 1S or (4D, 4F) $\rightarrow$ 2P.

6.3 Shift of Spectral Lines of Diatomic Polar Molecules in an Elliptically Polarized QEF

Consider a diatomic polar molecule in an elliptically polarized QEF. Nonperturbed levels $E_{vJ}^{(0)}$ are degenerate with respect to a projection M of the angular momentum (v and J are the vibrational and rotational quantum numbers).

Therefore the problem is to find the correct states of zero order (CSZO) $|vJ\alpha\rangle$ and then the corrections to $E_{vJ}^{(0)}$. We solve this problem by using the general results from Sect. 2.2.

Let a nonperturbed stationary Hamiltonian of a quantum system have the eigen functions $|n\alpha\rangle$ (index α denotes degenerate sublevels) so that $H|n\alpha\rangle = E_n^{(0)}|n\alpha\rangle$. The aim is to find QSs of such a system under the influence of a perturbation having the form

$$W(t) = V\cos\omega t + U\sin\omega t \qquad (6.3.1)$$

(V and U operators act on the coordinates).

At $V, U \to 0$ degenerate QSs tend to certain linear combinations of the functions $|n\alpha\rangle$ which are determined by the type of perturbation. Recall that in the known case where there is only a stationary perturbation V, the CSZO will be not the arbitrary WFs $|n\alpha\rangle$ but only those for which $\langle n\alpha|V|n\beta\rangle \propto \delta_{\alpha\beta}$.

The formalism from Sect. 2.2 for a perturbation of the type (6.3.1) gives the following result. For the functions $|n\alpha\rangle$ to be CSZO they must diagonalize the energy shift operator ε,

$$\varepsilon = \sum_{n'\alpha'}(V|n'\alpha'\rangle\langle n'\alpha'|V + U|n'\alpha'\rangle\langle n'\alpha'|U)\omega_n^{n'}$$

$$+ \,\mathrm{i}(V|n'\alpha'\rangle\langle n'\alpha'|U - U|n'\alpha'\rangle\langle n'\alpha'|V)\omega \times \left\{2\hbar[\omega^2 - (\omega_n^{n'})^2]\right\}^{-1},$$

$$(6.3.2)$$

where $\omega_n^{n'} \equiv (E_{n'}^{(0)} - E_n^{(0)})/\hbar$. On the basis of such CSZO, the diagonal elements of the operator ε will yield the needed corrections $\varepsilon_{n\alpha}$ to the nonperturbed energies $E_n^{(0)}$. The applicability criterion is reduced to

$$|\varepsilon_{n\alpha}| \ll \min(\omega, \omega_n^{n'}). \qquad (6.3.3)$$

Here we enumerate the analytical results for the corrections ε_α^{vJ} to the levels $E_{vJ}^{(0)}$ of diatomic molecules which have been obtained using this algorithm for the perturbation

$$W(t) = -E_0(1 + \xi^2)^{-1/2}\mu(x)(e_1\cos\omega t + \xi e_2\sin\omega t). \qquad (6.3.4)$$

As before E_0 is the QEF amplitude, ξ is the degree of QEF ellipticity, e_1 and e_2 are the orthogonal basis vectors, and $\mu(x)-$ is the molecular dipole moment [for the lowest vibrational states $\mu(x) \approx \mu_0 + \mu_1(x)$].

For any degree of ellipticity ξ, the energy corrections ε_α^{vJ} for $J = 1, 2$ are

$$\varepsilon_1^{v0} = \frac{2}{3}a_v; \quad \varepsilon_1^{v1} = \frac{4}{5}a_v, \quad \varepsilon_{2,3}^{v1} = \frac{3}{5}a_v$$

$$\pm\left[\left(\frac{\xi b_v}{1+\xi^2}\right)^2 + \left(\frac{1-\xi^2}{1+\xi^2}\cdot\frac{a_v}{5}\right)^2\right]^{1/2};$$

$$a_v \equiv E_0^2\left\{\left[\frac{\mu_0^2}{\hbar\omega^2} + \frac{\mu_1^2(\omega^2 + \omega_0^2)}{m\omega_0(\omega^2 - \omega_0^2)^2}\left(v + \frac{1}{2}\right)\right]\pi c B - \frac{\mu_1^2}{4m(\omega^2 - \omega_0^2)}\right\};$$

$$b_v \equiv E_0^2 \left\{ \left[\frac{\mu_0^2}{\hbar\omega^2} + \frac{\mu_1^2\omega(\omega^2 + 3\omega_0^2)}{m\omega_0(\omega^2 - \omega_0^2)^3} \left(v + \frac{1}{2} \right) \right] 8\pi^2 c^2 B^2 + \frac{2\pi c B \mu_1^2 \omega}{m(\omega^2 - \omega_0^2)^2} \right\},$$

(6.3.5)

where B (cm^{-1}) is the rotational constant. As it was expected, in the elliptically polarized field (at $\xi \neq 0$) the degeneracy of levels with $J \geqslant 1$ is totally lifted.

In circular polarization ($\xi^2 = 1$) the matrix $\varepsilon_{vJM}^{vJM'}$ is diagonal, so that for each J the energy corrections are

$$\varepsilon_M^{vJ} = a_v(3J^2 + 3J - 2 - M^2)(2J - 1)^{-1}(2J + 3)^{-1} \pm b_v M/2,$$

(6.3.6)

where the plus sign corresponds to the right-hand polarized and the minus sign to the left-hand polarized wave.

In linear polarization ($\xi = 0$) it is best to use a coordinate system with the polar axis along the EF where the perturbation is

$$V(t) = -(\mu_0 + \mu_1 x)E_0(\cos\theta) \cos\omega t.$$

(6.3.7)

In this basis the matrix $\varepsilon_{vJM}^{vJM'}$ is diagonal. For any J, the energy corrections will be

$$\varepsilon_M^{vJ} = 2a_v(J^2 + J - 1 + M^2)(2J - 1)^{-1}(2J + 3)^{-1}.$$

(6.3.8)

The problem with perturbation (6.3.7) was solved in [6.8] by introducing the Kapitza effective potential [6.9]. First the authors of [6.8] considered a rigid rotator which corresponds to the limit $\omega_0 \to \infty$. The result obtained at this limit is easily reproducible if matrix elements of the operator ε from (6.3.2) (for $U = 0$) are calculated in the subspace of the fixed vibrational quantum number v.

The effective potential could not be directly applied to the vibrating rotator, since the frequency ω is of the same order as ω_0 and the perturbation cannot be treated as rapidly varying. Hence the authors of [6.8] subdivided the system into rapid and slow subsystems, so that for the latter a differential equation was obtained:

$$[(-\partial^2/\partial y^2 + \omega_0^2 y^2)/2 - fy\cos\omega t]\psi(y, t) = i\partial\psi/\partial t,$$

(6.3.9)

where ω, ω_0, f are constants. They took the solution of equation (6.3.9) from [6.10].

However, there is a mistake in [6.10], and thus the result from [6.8] for the vibrating rotator is wrong. If within the approach used in [6.8] [which leads to more complicated calculations than when using the operator from (6.3.2)] all the operations with the correct solution of (6.3.9) are performed, then the result (6.3.8) is obtained with a_v from (6.3.5).

The shifts ε_α^{vJ} we obtained, as well as dynamical polarizabilities (connected with shifts ε by the known relation $\alpha = -4\varepsilon/E_0^2$), are determined mainly by the value of a_v. We evaluate it numerically for an HCl molecule in a CO_2 laser field. Substituting into (6.3.7) $\omega = 2 \times 10^{14}$ s^{-1}, $\omega_0 = 5.6 \times 10^{14}$ s^{-1}, $B = 10.6$ cm^{-1}, $\mu_0 = 1.09$ D, $\mu_1 = 0.949 \times 10^8$ D/cm and $E_0 = 10^6$ V/cm we obtain

$$a_v/2\pi\hbar c = [2 \times 10^{-3} + 2.5 \times 10^{-6}(v + 1/2)] \text{ cm}^{-1}.$$

Such shifts or splitting of rotational levels ($\sim 10^{-3}$ cm^{-1}) could be seen in rotational spectra where the Doppler width is $\sim 10^{-4}$ cm^{-1}. Experimental observation of this phenomenon would be of both general and practical interest: it would prove that from the shifts of molecular lines, using (6.3.5, 6,8) QEF parameters could be measured in low-temperature plasmas.

7 Applications of the Theory to Experimental Plasma Diagnostics

Methods of plasma diagnostics based on an analysis of SL profiles have, as is known, several important advantages. In particular, they do not affect the plasma parameters and do not depend upon the choice of the model for the plasma state. In Chaps. 3–6 we have presented a theory of SL shapes in plasmas containing QEFs. As was mentioned in Chap. 1, in practice, one has to solve every time the inverse (with respect to that discussed in Chaps. 3–6) problem, i.e., to determine the parameters of the QEF and the plasma media from the observed SL profiles. "Single-valued" algorithms for solving such problems for various plasma media do not exist (at least, at present). However, some general recommendations are pointed out in Sect. 7.1. The rest of this chapter is devoted to a discussion of experimental spectroscopic investigations carried out in various areas of plasma research.

7.1 Preliminary Remarks

Some effects of the influence of QEFs can be singled out by considerably simplifying a real situation. For non-Coulomb radiators, such an effect is the appearance of few satellites. In relatively rare plasmas (at $N_e \ll N_e^{cr}$) the nearest satellites appear at distances $\pm\omega$ from the dipole-forbidden SL (ω is the frequency of the QEF) and in denser plasmas ($N_e \gtrsim N_e^{cr}$) satellites may also appear at $\pm\omega$ from dipole-allowed SL (the latter is due to the enhanced influence of the quasistatic EF F of a plasma medium). The values of N_e^{cr} for the most often used SLs of helium are $N_e^{cr} \sim 10^{14}$ cm^{-3} for He I 447.2 nm and 492.2 nm; $N_e^{cr} \sim 10^{16}$ cm^{-3} for He I 501.6 nm and 667.8 nm.

For Coulomb radiators the situation is more complicated. On one hand the influence of a QEF without allowance for quasistatic fields F must lead to the appearance of satellites at $\pm\omega$, $\pm2\omega$, $\pm3\omega$, ... away from an unperturbed SL position. In the case of a relatively strong QEF [where $\omega \ll (Zm_e e)^{-1}(n^2 - n'^2)\hbar E_0 \equiv (\Delta\omega)_{E_0}$] far satellites have a significant statistical weight, so that even under a small additional broadening with a characteristic scale $\Delta\omega_{1/2}^{add} \gtrsim \omega$ (e.g., for Doppler or instrumental broadening) a satellite envelope with a characteristic width $(\Delta\omega)_{E_0}$ will be observed. On the other hand with allowance for quasistatic fields F the influence of a QEF should be the appearance of dips (or depressions). Both satellites and dips may appear in approximately the same range of wavelengths $\Delta\lambda \gtrsim \omega\lambda_0^2/2\pi c$. The question arises: what will be observed on experimental hydrogen-like SL in plasmas containing QEFs (e.g. Langmuir oscillations) – satellites, dips or both?

It should be emphasized that for each Stark component the inequality

$$X_{\alpha\beta}\varepsilon/Z \gg 1 \quad [\varepsilon \equiv 3\hbar E_0/2m_e e\omega, \; X_{\alpha\beta} \equiv n(n_1 - n_2) - n'(n'_1 - n'_2)],$$

determining the multisatellite (i.e. strong modulation) case is incompatible with the condition under which dips appear (4.2.22). Therefore the above question is, in fact, related to the opposite case $X_{\alpha\beta}\varepsilon/Z \ll 1$. In this case SL profiles of the Blochinzew type (Sect. 3.1.1) and of the Lifshitz type (Sect. 3.2.1) nearly coincide:

$$I_B(\Delta\omega) \approx I_L(\Delta\omega) \approx [1 - (X_{\alpha\beta}\varepsilon)^2/2Z^2]\delta(\Delta\omega)$$
$$+ X_{\alpha\beta}\varepsilon/(2Z^2)[\delta(\Delta\omega + \omega) + \delta(\Delta\omega - \omega)]. \tag{7.1.1}$$

The halfwidth of the entire SL does not actually change; only a local "zigzag" of intensity (near frequencies $\Delta\omega = \pm\omega$) arises:

$$\Delta I_{\text{sat}}/I \sim (n^2 - n'^2)^2\varepsilon^2/Z^2 \ll 1. \tag{7.1.2}$$

Near a dip the relative zigzag of intensity is of the order of $(\Delta I_{\text{dip}}/I)_\alpha \sim n/|X_{\alpha\beta}|$. For a given hydrogen-like SL the most pronounced dip is located on the profile of the most intense Stark component.

The question of which of the Stark components of a hydrogen-like SL is the most intense is of independent theoretical interest. As our analysis of the *Underhill - Waddell* tables [7.1] has shown, this question may be answered in a general form. For SL containing no central components ($n + n'$ is even) the most intense component is that with the quantum numbers

$$n_1 = (n - n')/2, \quad n_2 = (n - n' - 2)/2, \quad m = n';$$
$$n'_1 = n'_2 = 0, \quad m' = n' = 1. \tag{7.1.3}$$

For this component one has $X_{\alpha\beta} = n$, so that $\omega_\alpha^{\text{dip}} = \omega$. For head lines of spectral series ($n' = n - 1$) the most intense of the lateral components is characterized by the quantum numbers

$$n_1 = 1, \quad n_2 = 0, \quad m = n - 2; \quad n'_1 = 1, \quad n'_2 = 0, \quad m' = n' - 2. \tag{7.1.4}$$

For this component one also has $X_{\alpha\beta} = n$, so that $\omega_\alpha^{\text{dip}} = \omega$. Lastly, for other hydrogen-like SL ($n + n'$ is odd, $n \geqslant n' + 3$) the most intense of the lateral components has the quantum numbers

$$n_1 = (n - n' + 1)/2, \quad n_2 = (n - n' - 3)/2, \quad m = n';$$
$$n'_1 = n'_2 = 0, \quad m' = n' - 1. \tag{7.1.5}$$

For this component $X_{\alpha\beta} = 2n$, so that $\omega_\alpha^{\text{dip}} = 2\omega$.

Thus for the most pronounced dip on a hydrogen-like SL profile we find

$$|\Delta I_{\text{dip}}|/I \sim n/|X_{\alpha\beta}| \sim 1. \tag{7.1.6}$$

Using (7.1.2) we obtain

$$\Delta I_{\text{sat}}/|\Delta I_{\text{dip}}| \sim (n^2 - n'^2)^2\varepsilon^2/Z^2 \ll 1. \tag{7.1.7}$$

Hence the features on experimental hydrogen SLs in plasmas containing relatively weak QEFs [i.e. under the condition of (4.2.22)] should be mostly caused by the dip effect and not by the satellite effect.

Note that to detect dips higher spectral resolution is required than for observation of a satellite envelope (or even individual satellites) since usually the dip halfwidth is $\Delta\omega_{1/2}^{\text{dip}} \ll \omega$. When the scale $\Delta\omega_{1/2}^{\text{add}}$ of competing broadening mechanisms is sufficiently large ($\Delta\omega_{1/2}^{\text{add}} \gtrsim \omega \gg \Delta\omega_{1/2}^{\text{dip}}$) the dip cannot be seen.

In conclusion we would like to summarize our recommendations to experimentalists. Before measuring the QEF parameters in a plasma it is desirable to know the electron density $N_e = Z_i N_i$, the temperatures T_e, T_i, the charge of the perturbing ions Z_i and also to determine the scales of other competing mechanisms of broadening or splitting, such as Doppler, Zeeman, and Stark broadening by electrons and ions, and broadening due to self-absorption. The expected amplitude of QEFs can be estimated from the inequality $E_0 \lesssim [8\pi(N_e T_e + N_i T_i)]^{1/2}$. For non-Coulomb radiators these parameters allow one to choose SLs for which, first, the expected satellites might be sufficiently intense and, second, the competing broadening effect should not prevent detection of satellites. For Coulomb emitters a knowledge of these parameters allows the determination of hydrogen-like SLs for which the scale $(\delta\lambda)_{E_0}$ of the effects of QEFs (i.e. the broadening in the multisatellite case, the appearance of dips or nearest satellites in the opposite case) dominates over competing effects.

It is important to emphasize that spectroscopic diagnostics of QEF in plasmas is also possible when for the head lines of the hydrogen-like spectral series (L_α, L_β, H_α, H_β) the value of $(\delta\lambda)_{E_0}$ is small compared to $\Delta\lambda_{1/2}^{\text{add}} = \Delta\omega_{1/2}^{\text{add}}\lambda_0^2/2\pi c$ (e.g. in comparison with the Doppler broadening). In this case diagnostics can be performed when a polarizer is inserted into the optical system. Then, one or two intense (head) lines in two orthogonal linear polarizations I_1, I_2 should be examined [if the QEFs are anisotropic, the profiles $I_1(\Delta\lambda)$, $I_2(\Delta\lambda)$ may differ]. In another version the observations are carried out with unpolarized light but the profiles of several SLs for which $(\delta\lambda)_{E_0} \gtrsim \Delta\lambda_{1/2}^{\text{add}}$ are recorded. This version uses the fact that for more highly excited SLs the ratio $(\delta\lambda)_{E_0}/\Delta\lambda_{1/2}^{\text{add}}$ increases significantly and, as a rule, monotonically[1].

7.2 QEFs in θ-Pinches

7.2.1 QEFs Under a Magnetic Field Annihilation

Experiments at the "Dimpol" installation, a magnetic mirror trap described in [7.2] were reported in [7.3]. On an initial plasma created by a Penning-type discharge in a dc magnetic field an ac magnetic field of high amplitude

---

[1] Sometimes in practice one also finds a nonmonotonic dependence of the ratio $(\delta\lambda)_{E_0}/\Delta\lambda_{1/2}^{\text{add}}$ or $(\delta\lambda)_{E_0}$ on the principal quantum number. Such a situation is analyzed in Sect. 7.5.2.

was imposed. Thus an opposing magnetic fields configuration arose in a θ-pinch geometry. The investigations were performed with hydrogen at a pressure $P_0 \approx 0.5$ Pa, a dc magnetic field $B_0 = 0.07$ T and an ac magnetic field $B = 0.28-0.42$ T. In the preliminary study [7.2] the existence of low-frequency turbulent BEF distributed anisotropically was revealed, with a dominant BEF component orthogonal to the magnetic field

$$F_{\|} \equiv F_z \approx 8 \text{ kV/cm}, \quad F_{\perp} \equiv (F_{\varphi}^2 + F_r^2)^{1/2} \approx 30 \text{ kV/cm}, \quad F_{\varphi} \approx F_r$$

(unit vectors of the cylindrical coordinate system are indicated in Fig. 7.1). However, it was not clear whether QEFs (in particular, Langmuir oscillations) arose.

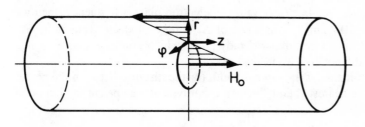

Fig. 7.1. Unit vectors in a cylindrical coordinate system used to analyze the distributions of EFs in a plasma [7.3]

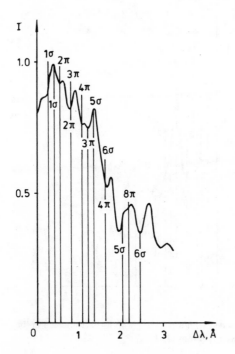

Fig. 7.2. Profile of the H_α line recorded in longitudinal observation without a polarizer. The *vertical line segments* indicate the theoretically expected dip positions. From [7.3]

Figure 7.2 shows the profile of the H_α line, recorded longitudinally without a polarizer. The vertical line segments indicate the theoretically expected dip positions. It is seen that the measured depressions coincide exactly with the positions of nine (out of 12 expected) dips and correspond to the distances $\lambda_\omega/3$, $2\lambda_\omega/3$, λ_ω, $4\lambda_\omega/3$, $3\lambda_\omega/2$, $2\lambda_\omega$, $5\lambda_\omega/2$, $8\lambda_\omega/3$ where $\lambda_\omega = \omega\lambda_0^2/2\pi c$. The three remaining dips should appear at $\lambda_\omega/2 = \lambda_{100}^{1\sigma}$, $5\lambda_\omega/3 = \lambda_{101}^{5\sigma}$, $4\lambda_\omega = \lambda_{010}^{8\pi}$. Here the notation $\lambda_{101}^{5\sigma}$ means that the dip is formed on the profile of a σ-component with $X_{\alpha\beta} \equiv n_\alpha(n_1 - n_2)_\alpha - n_\beta(n_1 - n_2)_\beta = 5$ due to a resonance with the upper multiplet, to which the initial (for the 5σ - component) sublevel with quantum numbers $n_1 = 1$, $n_2 = 0$, $|m| = 1$ belongs. The dips at $\lambda_\omega/2$ and $5\lambda_\omega/3$ have apparently merged with the dips at $2\lambda_\omega/3$ and $3\lambda_\omega/2$ respectively; the wavelength $4\lambda_\omega$ corresponds to a remote wing outside the measured band. From the indicated structure one finds $\lambda_\omega \approx 0.87$ Å. If we assume that the QEF frequency $\omega = \omega_{pe}$, where $\omega_{pe} = (4\pi e^2 N_e/m_e)$ is the electron plasma frequency, then we find $N_e = 4.5 \times 10^{13}$ cm^{-3}. This value of N_e agrees satisfactorily with the N_e value measured in the preliminary study [7.3], which confirms that $\omega \approx \omega_{pe}$ (below we use the notation λ_p instead of λ_ω).

The measurements of the dip halfwidth (or the distance between the center of the dip and the nearest "bump", which is equivalent) can be carried out quite

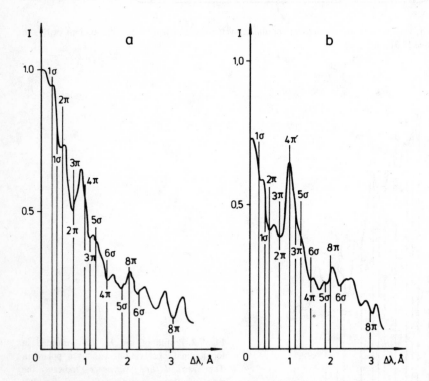

Fig. 7.3. Polarization contours of the H_α line obtained with a polarizer with its axis oriented in the z plane (**a**) and φ plane (**b**) in a transverse observation. From [7.3]

accurately on six dips of the H_α profile shown in Fig. 7.2: $(\Delta\lambda)_{2/3} \approx 0.11$ Å, $(\Delta\lambda)_1 \approx 0.15$ Å, $(\Delta\lambda)_{3/2} \approx 0.11$ Å, $(\Delta\lambda)_2 \approx 0.14$ Å, $(\Delta\lambda)_{5/2} \approx 0.16$ Å, $(\Delta\lambda)_3 \approx 0.22$ Å (the subscript on $\Delta\lambda$ indicates the distance from the dip to the line center in units of λ_p). Substituting the measured halfwidths into (4.2.46), we obtain the amplitude $E_0 = 4.3 \pm 0.5$ kV/cm.

To determine the directivity diagram of Langmuir waves we introduced a polarizer into the optical system, which made it possible, in transverse observation, to separate the z and φ polarizations and in longitudinal observation to separate the r and φ polarizations (Fig. 7.1). Figure 7.3 shows typical z and φ polarization contours obtained in transverse observation, while Fig. 7.4 shows r and φ polarization contours obtained in longitudinal observation. The value of λ_p determined by measuring the distance between the dips is the same for all four contours and equals $\lambda_p \approx 0.76$ Å, which corresponds to a plasma concentration of $N_e \approx 3.4 \times 10^{13}$ cm^{-3}. The slight decrease of the concentration from the previously measured $N_e \approx 4.5 \times 10^{13}$ cm^{-3}, achieved by decreasing the hydrogen pressure, was required for farthest the dip in the line wing to fall within the measured band of 3.6 Å.

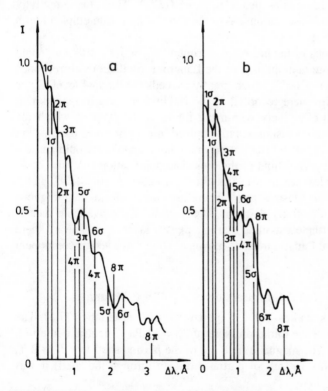

Fig. 7.4. Polarization contours of the H_α line obtained with a polarizer axis in the r plane (**a**) and φ plane (**b**) in longitudinal observation. From [7.3]

When comparing dips on the polarization profiles, it must be taken into account that, according to the nonadiabatic theory, the dip should be flanked by two "bumps", which compensate for the normalization. However, when two dips are in close proximity to each other usually only one bump can be observed between them, rather than two. Thus, close proximity of two or more dips distorts their halfwidths. This effect can exceed the polarization difference between the half-width of the dip on the r and φ profiles. Therefore the most reliable results are obtained from a polarization analysis of well-isolated dips. On the H_α profile, the most isolated of all the dips is at $4\lambda_p = \lambda_{010}^{8\pi}$. Between this dip and the nearest dip of wavelength $\lambda_{001}^{6\sigma} = 3\lambda_p$ two bumps on each of the polarization profiles are observed (Fig. 7.3, 4). To some degree, the dips at λ_p, $2\lambda_p$, and $3\lambda_p$ can also be regarded as isolated. A statistical comparison of the halfwidths of these dips on the z and φ profiles, carried out for different pairs of profiles obtained in transverse observation revealed no noticeable polarization difference between the halfwidths: $(\Delta\lambda_{1/2}^{dip})_{z\pi} \approx (\Delta\lambda_{1/2}^{dip})_{\varphi\pi}$, $(\Delta\lambda_{1/2}^{dip})_{z\sigma} \approx (\Delta\lambda_{1/2}^{dip})_{\varphi\sigma}$ and consequently $\langle E_3^2 \rangle \approx \langle E_\varphi^2 \rangle$. A comparison of the halfwidths of the same dips on the r and φ profiles obtained in longitudinal observation shows that $(\Delta\lambda_{1/2}^{dip})_{r\pi} > (\Delta\lambda_{1/2}^{dip})_{\varphi\pi}$, $(\Delta\lambda_{1/2}^{dip})_{r\sigma} < (\Delta\lambda_{1/2}^{dip})_{\varphi\sigma}$. In accordance with the theoretical results [7.4] this means that $(\langle E_r^2 \rangle) < (\langle E_\varphi^2 \rangle)$. Thus, the directivity diagram of the Langmuir oscillations takes the form of an oblate ellipsoid with the symmetry axis along r.

Under the conditions of the present experiment, $N_e = 3.4 \times 10^{13}$ cm^{-3} and $T_e \sim 10^3$ eV, the measured amplitude of the Langmuir oscillations corresponds to a level $E_0^2/(8\pi N_e T_e) \sim 10^{-4}$, which greatly exceeds the thermal level. Since clearly pronounced dips were recorded in the H_α line profiles, it appears that the QEF is developed effectively within the limits of the layer in which the plasma concentration varies insignificantly. Indeed, the positions of the dips are proportional to $N_e^{1/2}$, so that substantial plasma inhomogeneities would cause a "smearing" and in fact a vanishing of the dips. The polarization analysis shows that the Langmuir fields are oriented in the "current-magnetic field" plane, which is tangent to the plasma layer. Therefore, although an unambiguous interpretation of this directivity diagram on the basis of the available data is difficult, one can nevertheless assume that in a plasma layer of thickness δ there develops a fundamental Langmuir mode having a radial wave vector component $k_r \sim \delta^{-1}$.[2]

7.2.2 QEFs Under a Rapid Compression of a θ-Pinch Plasma

Deuterium plasma was studied at the θ-pinch "UTRO" installation [7.5]. The electron concentration at the initial pressure of $P_0 = 13$ Pa of neutral deuterium was $N_e \approx 10^{16}$ cm^{-3}. The magnetic field out of the plasma shell was $B \approx 2$ T, on the chamber axis (as a result of partial diffusion through the shell) it was ≈ 0.4 T.

---

[2] This was pointed out to us by L.I. Rudakov.

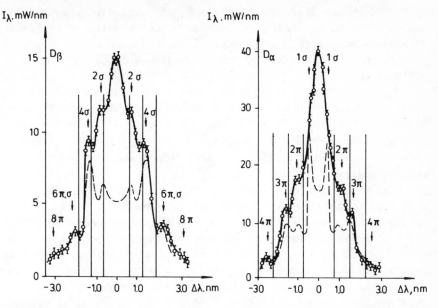

Fig. 7.5. Experimental profiles of the deuterium lines D_α and D_β *(solid lines)*. *Dashed lines*: theoretical profiles calculated using the adiabatic (Blochinzew type) approximation for the action of QEFs of low-hybrid oscillations *(arrows* show the positions of Stark component maxima). All data From [7.5]. *Vertical lines*: theoretically predicted positions of dips calculated using our assumption of nonadiabatic joint action of QEFs of Langmuir oscillations and quasistatic EFs of plasma.

The profiles of D_α and D_β lines, recorded at the stage of rapid compression ($t \approx 0.4$ μs) in two discharges with almost the same parameters, are shown in Fig. 7.5. The authors of [7.5] have interpreted the observed features as Blochinzew type satellites caused by regular low-hybrid oscillations with the amplitude $E_0 = 220$–250 kV/cm (dashed lines).

They explain the discrepancy between the measured and calculated intensities by the presence of additional emission from "cold" regions where there are no low-hybrid QEFs. According to the estimations of [7.5] the neutral deuterium concentrations in the states $n = 3$ and $n = 4$ in the cold and hot (containing QEFs) regions are approximately equal $n_D^{(3)} \sim n_D^{(4)} \sim (10^{10}$–$10^{11})$ cm^{-3} but in the $n = 2$ state the cold deuterium atoms dominate and their concentration is $n_D^{(2)} \sim (5$–$10) \times 10^{12}\tau_0$ cm^{-3}. Here τ_0 is the optical thickness in the center of the D_α line, which in [7.5] has been assumed to be equal to unity.

We believe that the interpretation in [7.5] is incorrect. At $B = 2$ T the low-hybrid oscillation frequency is equal to $\omega = eB/[c(m_e M_i)^{1/2}] \approx 5.8 \times 10^{10}$ s^{-1} (M_i is ion mass), so that at $E_0 = 220$ kV/cm the dimensionless parameter from Sect. 3.1.1 is equal to $\varepsilon = 3\hbar E_0/2m_e e\omega \approx 46$. This means that both for D_α and for D_β there is strong modulation of the emitted light wave. In this case, with (3.1.12) it is easy to obtain that for the theoretical profile of the absorption coefficient of the D_α line the ratio of the intensity at the maximum of the profile of the 3π -component (the most intense of the lateral components) to the central component intensity is equal to $I_{3\pi}/I_{0\sigma} \approx 0.014$. But in the experimental profile the corresponding ratio is 20 times greater: $(I_{3\pi}/I_{0\sigma})_{\exp} \approx 0.3$. In addition, in [7.5] it is claimed that not less than half of the intensity in the center of the D_{α} line is emitted by the cold plasma layer. Consequently the cold layer reduces the intensity in the center of the D_{α} profile emitted by the hot region by not less than 40 times, so that the optical thickness in the center of D_{α} should be $\tau_0 \gtrsim \ln 40 \approx 3.7$. Thus the deuterium atom concentration in the $n = 2$ level is equal to $n_{\mathrm{D}}^{(2)} \approx (2\text{--}4) \times 10^{13}$ cm^{-3}. However, it is doubtful that any model of plasma equilibrium could explain such a striking difference of population ratios: $n_{\mathrm{D}}^{(3)}/n_{\mathrm{D}}^{(2)} \sim 10^{-3}$, $n_{\mathrm{D}}^{(4)}/n_{\mathrm{D}}^{(3)} \sim 1$.

Even more significant are the following considerations. Suppose, as was done in [7.5], that the observed QEFs are really low-hybrid oscillations with an amplitude $E_0 = 220$ kV/cm (so that $\varepsilon \approx 46$, as mentioned above). Then in the D_{α} profile the ratios of the maxima of different components should be, (according to (3.1.12), $I_{2\pi}/I_{3\pi} = 0.41$ and $I_{4\pi}/I_{3\pi} = 0.6$. But the corresponding experimental ratios are equal to $(I_{2\pi}/I_{3\pi})_{\exp} \approx 1.4$ and $(I_{4\pi}/I_{3\pi})_{\exp} \approx 0.3$, so that the latter is twice as much as the true ratio and the former three times as much and gives a result that is even qualitatively wrong: $I_{2\pi} > I_{3\pi}$ instead of $I_{2\pi} < I_{3\pi}$. For the D_{β} line the situation is analogous. Indeed according to the theory, under the conditions of the experiment it should be that $I_{2\sigma}/I_{4\sigma} = 0.28$ and $I_{6\pi,\sigma}/I_{4\sigma} = 0.63$. But the experimental values are $(I_{2\sigma}/I_{4\sigma})_{\exp} \approx 1.2$ and $(I_{6\pi,\sigma}/I_{4\sigma}) \approx 0.4$, so that the former is more than four times as large as the true ratio and gives a qualitatively wrong result: $I_{2\sigma} > I_{4\sigma}$ instead of $I_{2\sigma} < I_{4\sigma}$.

Even if we suppose that the intensity of components near the line center is distorted by the cold plasma radiation and one should rely on the far components, then for the D_{α} line, e.g., we obtain: $(I_{8\pi}/I_{6\pi,\sigma})_{\exp} \approx 0.5$ instead of the true ratio $I_{8\pi}/I_{6\pi,\sigma} = 0.84$. Thus even for far components of D_{α}, D_{β} lines the ratio of intensities is almost twice as much as the true one. Summarizing all these arguments one may assert that the interpretation given in [7.5] is unlikely.

Let us try to identify the observed features of D_{α}, D_{β} profiles not via satellites but via dips. The possible dip positions are shown in Fig. 7.5 by the vertical lines at $\pm\lambda_{\omega}/2$, $\pm\lambda_{\omega}$, $\pm3\lambda_{\omega}/2$ from the D_{α} line center and at $\pm\lambda_{\omega}$, $\pm2\lambda_{\omega}$, $\pm3\lambda_{\omega}$ from the D_{β} line center (recall that $\lambda_{\omega} \equiv \omega\lambda_0^2/(2\pi c)$, where ω is the QEF frequency and λ_0 the unperturbed wavelength of the corresponding SL). Relative separations of dips from the center do not depend on λ_{ω}; the absolute scale λ_{ω} for each profile was chosen from the condition of the best fit of observed and theoretical dip positions. These scales are equal to $\lambda_{\omega} \approx 1.5$ nm for D_{α} and $\lambda_{\omega} \approx 0.65$ nm for D_{β}. Consequently the QEF frequency determined by the D_{α} line is equal to $\omega \approx 6.5 \times 10^{12}$ s^{-1} and by the D_{β} line it is $\omega \approx 5.2 \times 10^{12}$ s^{-1}, which agree with each other to within 10% error.

If one identifies the obtained value of the QEF frequency with the frequency of Langmuir oscillations, then for the electron concentration one obtains $N_e = (1.1 \pm 0.25) \times 10^{16}$ cm^{-3}, which agrees well with the known (measured by another method) value, $N_e \approx 10^{16}$ cm^{-3}. Thus it seems that in [7.5] at the stage of rapid compression of the θ-pinch the Langmuir (and not low-hybrid) oscillations were developed in the plasma layer.

7.3 QEFs in a Z-Pinch

The spectroscopy of a deuterium plasma with QEF in a Z-pinch configuration has also been investigated [7.6]. The maximum discharge current reached $I_{max} = 350$–400 kA, with $I(t_I) \approx 200$ kA and $I(t_{II}) \approx 300$ kA at the instants of the first and the second singularities, respectively. The current rise time to the maximum was 4.5 μs. The maximum value of the time derivative of the current was in this case $\dot{I}_{max} \approx 4 \times 10^{11}$ A/s.

Typical profiles of the Balmer lines D_α, D_β, and D_γ recorded during different phases of the discharge are shown in Figs. 7.6–9. The principal feature of all the profiles is the richness of the structure: in each wing several intensity dips are observed, sometimes almost decreasing to zero. Let us see what effects can account for the observed singularities.

The magnetic field of the discharge current is $H = 2I/cr \leqslant 6 \times 10^6$ A/m, since the observed radius of the pinch is $r \gtrsim 1$ cm. In such fields, the quadratic Zeeman effect on the lines D_α, D_β, and D_γ is small compared with the linear one, so that only the triplet magnetic splitting is possible. Consequently, the observed number of peaks (on the order of ten on the profiles of D_β and D_γ) cannot be attributed to the Zeeman effect.

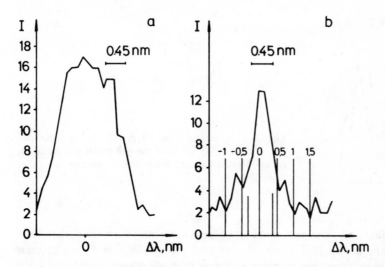

Fig. 7.6. Experimental profiles of the D_α line at a pressure $P_0 = 20$ Pa: (**a**) at the instant of the first negative spike of the current; (**b**) after the instant of the second negative spike. The *vertical line segments* indicate the theoretically expected positions of the dips. The segments are marked by the distances of the dips from the line center in units of λ_ω. From [7.6]

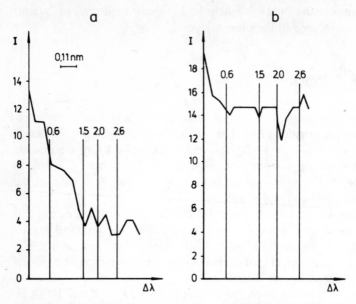

Fig. 7.7. Experimental profiles of the D_γ line at a pressure $P_0 = 20$ Pa: (**a**) prior to the instant of the first negative spike of the current; (**b**) at the instant of the first negative spike of the current. The *vertical lines* and the numbers above them as in Fig. 7.6. From [7.6]

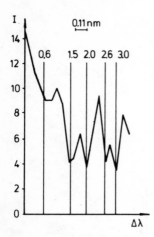

Fig. 7.8. Experimental profile of the D_γ line at a pressure $P_0 = 11$ Pa at the instant of the first negative spike of the current. Vertical lines and numbers above them as in Fig. 7.6. From [7.6]

To explain the observed number of peaks we might propose that low-frequency EFs are excited in the plasma, and the distribution $W(F)$ of their amplitudes is for some reason much narrower than a Rayleigh distribution, and the field F can be regarded as quasihomogeneous. But in this case the halfwidth of the Stark component $(\Delta\lambda_{1/2})_{\alpha\beta} = C_{\alpha\beta}\Delta F_{1/2}^{(W)}$ [$\Delta F_{1/2}^{(W)}$ is the halfwidth of the $W(F)$ distribution] is proportional to its distance from the center of the line.

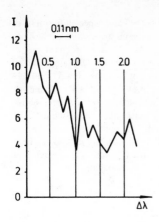

Fig. 7.9. Experimental profile of the D_β line at a pressure $P_0 = 20$ Pa at the instant of the first negative spike of the current. *Vertical lines* and numbers above them as in Fig. 7.6. From [7.6]

Therefore, farther into the wing the resolution of the individual components should become worse, in contradiction to the observed profiles.

For this reason, the observed profiles could also not be attributed to a combined Stark – Zeeman effect. In addition, this splitting depends substantially on the angle between H and F [7.7]. It is doubtful therefore that after averaging over the direction and magnitude of the vector F there could remain such a pronounced structure as observed in the present experiment.

Finally, we could suppose that QEF of frequency ω are excited in the plasma. If for some reason the principal (nonadiabatic) effect of their action could not appear, then their manifestation would reduce only to adiabatic satellites at $\lambda_0 + k\lambda_\omega$, where $k = \pm 1, \pm 2, \ldots$; $\lambda_\omega = \omega\lambda_0^2(2\pi c)^{-1}$. This, however, contradicts the observed peak distribution, which is not equidistant (especially on the D_γ line). We emphasize also that the adiabatic satellites cannot lead to such abrupt drops of intensity as can the nonadiabatic effect leading to the appearance of dips (Sect. 7.1).

Thus, it remains to check the possibility of attributing the observed structure of the profiles to the principal resonant effect of the action of QEF, as will be done below.

The spectral resolution in this experiments was 0.4Å for D_β, D_γ lines and 1.2 Å for the D_α line. When the spectral resolution is low some part of the dips cannot be recorded. It is convenient in this case to select beforehand a set of "reference dips" in the theoretical calculations, with the largest equivalent width w_ν, which can be defined by

$$w_\nu \equiv \int\limits_{\lambda_\nu - \delta\lambda_\nu}^{\lambda_\nu + \delta\lambda_\nu} d(\Delta\lambda)[S_0(\Delta\lambda) - S(\Delta\lambda)], \quad \delta\lambda_\nu = (\Delta\lambda_{1/2}^{\mathrm{dip}})_\nu, \tag{7.3.1}$$

where $S_0(\Delta\lambda)$ is the line profile unperturbed by QEFs. Calculation shows that

$$w_\nu \propto I_{\alpha\beta}n_\nu[n_\nu^2 - (n_1 - n_2)_\nu^2 - m_\nu^2 - 1]^{1/2}$$
$$\times [n_\alpha(n_1 - n_2)_\alpha - n_\beta(n_1 - n_2)_\beta]^{-1} \equiv f_\nu, \tag{7.3.2}$$

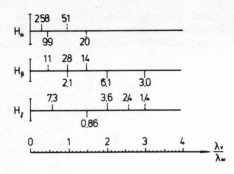

Fig. 7.10. Theoretically expected positions of the reference dips on the profiles of the lines D_α, D_β and D_γ. The dips produced due to the resonance between the QEF frequency ω and the splitting of the upper (lower) multiplet with $n_\alpha > 2$, i.e. α-group ($n_\beta = 2$, i.e. β-group) are indicated by *vertical lines* above (below) the abscissa axis. Alongside the lines the equivalent dip width w_ν is shown in relative units (7.3.1,2). *Solid lines* indicate dips for which $w_\nu \geqslant w_\nu^{\max}/2$, *dashed lines* the broadest of the remaining dips

where $I_{\alpha\beta}$ is the relative intensity of a Stark component and n_ν, $(n_1 - n_2)_\nu$, m_ν are the parabolic quantum numbers of the initial ($\nu = \alpha$) or final ($\nu = \beta$) states. The quantities f_ν, which are independent of the plasma parameters, determine the ratios of the equivalent widths within each of two groups (α or β) of dips: $w_{\alpha'}/w_\alpha = f_{\alpha'}/f_\alpha$, $w_{\beta'}/w_\beta = f_{\beta'}/f_\beta$.

Comparison of the relative distances of the observed dips from the line center with the theoretically expected relative positions of the reference dips (Fig. 7.10) shows good agreement for all the profiles of the lines D_α, D_β, and D_α at our disposal. In particular, for the typical profiles of D_α, D_β, and D_γ presented above this can be verified from an examination of Figs. 7.6–9, which show the theoretical positions of the dips (now on the absolute scale). This agreement for a large number of profiles of different Balmer lines can hardly be accidental, all the more so since on the D_γ profile the dips are not equidistant (their distances from the center are $3\lambda_\omega/5$, $3\lambda_\omega/2$, $2\lambda_\omega$, $13\lambda_\omega/5$).

From the absolute positions of the observed dips we obtain for each line the value of λ_ω, and then, assuming $\omega = \omega_{pe}$, the electron concentration N_e. It turns out that for profiles measured during the same discharge phase (e.g. at the instant of the first negative spike of the current), the experimentally measured λ_ω decreases from D_α to D_β and D_γ, with $\lambda_\omega \propto \lambda_0^2$, so that all three Balmer lines give the same value of N_e. This is an additional evidence of the reliable identification of the measured dips with the predicted ones.

Most experiments were performed at a molecular deuterium pressure of $P_0 = 20$ Pa. At the instant of the first negative spike t_1, the measurement of the plasma concentration via the λ_ω of the D_α, D_β and D_γ lines yields $N_e \approx 4 \times 10^{15} \text{cm}^{-3}$. As a control, experiments were also performed at a lower pressure $P_0 = 11$ Pa (Fig. 7.8). The relative positions of the dips on the profile of D_γ remained the same, and the absolute distances of the dips from the line center decreased, so that measurements relative to these lines yielded $N_e' \approx 2 \times 10^{15} \text{ cm}^{-3}$. The ratio $N_e/N_e' = 2$ is in agreement (within the limits of error) with the ratio $P_0/P_0' \approx 1.8$, and this again indicates that the experimental data have been correctly interpreted theoretically.

It is worth mentioning the work of Piel and Richter [7.8] in which it was shown that numerous weak peaks on the H_β line could also not be interpreted in the usual manner as being adiabatic satellites and were apparently molecular H_2 lines. Inasmuch as in the present experiment [7, 6] the distance from the center of the D_γ line to the observed structures depends on the pressure of the molecular deuterium ($\alpha P_0^{1/2}$), these structures are not molecular D_2 lines. This conclusion is also confirmed by a check with the table of D_2 lines [7.9], which show that there is no one-to-one correlation between the positions of the D_2 lines and the observed dips on the profiles of D_α, D_β and D_γ. We note also that the experiment in [7.8] was initiated by a proposal made in [7.10] that the structures in the profiles of He I can be attributed to the molecular lines of He_2. This idea, however, seems to be not confirmed [7.11].

It is of interest to compare the obtained value of the plasma concentration (on the periphery of the pinch) at the instant t_1, 4×10^{15} cm^{-3}, with the value obtained from an analysis of the halfwidths of the profiles. It was shown by us [7.12] that in theory, in the absence of oscillatory EF, the dominant broadening mechanism under the conditions of the present experiment is the impact action of the ions. In this case, the measured halfwidth of the line D_β, $\Delta\lambda_{1/2}(D_\beta) \approx$ 10 Å, at the time t_1 corresponds to $N_e > N_{min} \approx 1.0 \times 10^{16}$ cm^{-3} [7.12].

The substantially higher concentration obtained from $\Delta\lambda_{1/2}(D_\beta)$ indicates that an intense low-frequency BEF develops in the plasma; since the D_β line does not have a central Stark component, the quasistatic action of the BEF with a root-mean-square field F_0 increases $\Delta\lambda_{1/2}^{exp}(D_\beta)$ to [7.13]

$$\Delta\lambda_{1/2}^{theor}(D_\beta)[\text{Å}] \approx 0.2 F_0[\text{k V/cm}]. \tag{7.3.3}$$

Substituting $\Delta\lambda_{1/2}^{exp}(D_\beta) \approx 10$ Å, we get $F_0 \approx 50$ kV/cm.

The fact that the Stark broadening of the line is due to collective low-frequency fields and not to random thermal motion of the ions is also confirmed by polarization experiments. In these experiments the polaroid transmission axis was perpendicular to the axis of the pinch. An elementary analysis of the polarization of the light within the Stark profile shows that the ratio of the half widths $\Delta\lambda_{1/2}^{pol}$ and $\Delta\lambda_{1/2}^{unpol}$ of the polarized and unpolarized profiles of the line D_β should be

$$\eta \equiv \Delta\lambda_{1/2}^{pol}/\Delta\lambda_{1/2}^{unpol} \approx (F_\varphi^2 + F_r^2)/(F_\varphi^2 + F_r^2 + 2F_z^2) \tag{7.3.4}$$

(the z axis of the cylindrical coordinate system is directed along the current, and the φ axis along the magnetic field of the current). In particular, one should expect $\eta \approx 3^{-1/2}$ in the case $F_\varphi^2 \ll F_r^2 \approx F_z^2$ and $\eta \ll 3^{-1/2}$ in the case $F_z^2 \gg \max(F_\varphi^2, F_r^2)$.

A typical form of the polarization profile of the D_β line is shown in Fig. 7.11. The measured value $\Delta\lambda_{1/2}^{pol} \approx 6$ Å (i.e., $\eta \approx 0.6$) is evidence of anisotropy of the low-frequency fields and corresponds best to the case $F_\varphi^2 \ll F_r^2 \approx F_z^2$.

We emphasize that the values $F_0 \sim 50$ kV/cm and $N_e \sim 4 \times 10^{15}$ cm^{-3} measured at the time t_1 pertain to the periphery of the pinch (it follows from

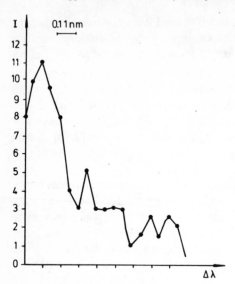

Fig. 7.11. Experimental polarization profile of the D_β line at a pressure $P_0 = 20$ Pa at the instant of the first negative spike of the current. The polaroid transmission axis is oriented perpendicular to the axis of the pinch [7.6]

laser-scattering experiments [7.14] that N_e at the center of the pinch can be larger by one and one-half or two orders of magnitude). Assuming that at the instant t_1 we have in accordance with [7.14] $T_i \gtrsim 30$ eV and $T_e \lesssim T_i$, we find that the turbulence level is $\lesssim 10^{-2}$.

Let us now analyze the halfwidth of the D_α line. At t_1 the value of $\Delta\lambda_{1/2}(D_\alpha)$ reaches 15.7 Å (Fig. 7.6, curve a), and then decreases with time (Fig. 7.6, curve b). We show first that the "laminar" (unrelated to the turbulence) broadening mechanisms in a plasma with parameters $T_a(t_1) \approx T_i(t_1) \geqslant 30$ eV and $N_e \approx 4 \times 10^{15}$ cm^{-3} cannot explain the measured value of $\Delta\lambda_{1/2}(D_\alpha)$.

For the D_α line, the ion impact halfwidth is $(\Delta\lambda_{1/2})_{\text{imp}}^{(i)} < 4$ Å. The Doppler halfwidth of D_α can reach 15.7 Å only at $T_a \approx 2$ keV, which exceeds the upper bound $T_a \lesssim 10^2$ eV obtained in the experiments [7.14]. The contribution of the remaining broadening mechanisms in a laminar plasma having the indicated parameters can be neglected.

It was shown in [7.12] that the "anomalous" value of $\Delta\lambda_{1/2}(D_\alpha)$ also cannot be attributed to self-absorption. Thus, to explain the value of $\Delta\lambda_{1/2}(D_\alpha)$ we must resort to "turbulent" broadening mechanisms, among which for the D_α line, the most important should be the broadening by Langmuir waves [7.15]. For the D_α line the corresponding theoretical halfwidth becomes $\Delta\lambda_{1/2}[\text{Å}] \lesssim 5 \times 10^4 (\gamma/\omega_{\text{pe}})[N_e, \text{cm}^{-3}]^{-1/2} E_0$ [kV/cm] so that at the values of N_e and E_0 indicated above we have $\Delta\lambda_{1/2} \lesssim 5\gamma/\omega_{\text{pe}} < 5$ Å (here γ is the halfwidth of the frequency spectrum of Langmuir waves). This formula, however, was obtained under the assumption $F \gg E_{\text{eff}} = E_0/\sqrt{2}$. In the case considered, most atoms radiate under conditions $F \lesssim E_{\text{eff}}$. The halfwidth of the line can then reach the value

$$(\Delta\lambda_{1/2})_{\text{max}} \approx \omega_{\text{pe}}\lambda_0^2/(\pi c). \tag{7.3.5}$$

Substituting in (7.3.5) the measured $\Delta\lambda_{1/2}(D_\alpha)$ of 15.7 Å, we get ω_{pe} and the electron concentration. The value $N_e \approx 4 \times 10^{15}$ cm^{-3} obtained in this manner agrees with the value obtained above by measuring the positions of the dips on the D_α, D_β and D_γ profiles. This attests once more to the adequacy of the theoretical interpretation of the experimental data.

The decrease of $\Delta\lambda_{1/2}(D_\alpha)$ after the time t_I can be attributed to the increase of the ratio F_0/E_0, as a result of which the fraction of the atoms that radiate under conditions $F \lesssim E_{eff}$ is decreased.

We now summarize the arguments supporting the described interpretation of the entire set of experimental profiles D_α, D_β, D_γ.

The QEF frequency determined from the dip positions is equal to $\omega \approx 3.6 \times 10^{12}$ s^{-1}. Strictly speaking, we may further only assume that $\omega \leqslant \omega_{pe}$ and determine only the lower limit of concentration $N_e \geqslant N_e^{min} = 4 \times 10^{15}$ cm^{-3}. The upper limit of N_e seems to be arbitrary at the first glance. However, the polarization measurements of the D_β halfwidth give $\Delta\lambda_{1/2}^{pol} \approx 6$ Å (and the unpolarized value is $\Delta\lambda_{1/2}^{unpol} \approx 10$ Å). Assuming that the value of 6 Å is entirely caused by the "laminar" mechanisms of Stark broadening (in our case – by the ion impact broadening) we find the upper limit of density: $N_e \leqslant N_e^{max} \approx 6 \times 10^{15}$ cm^{-3}. Thus the limits on the density N_e and correspondingly the frequency ω_{pe} are rather reliable: 3.6×10^{12} s$^{-1} \leqslant \omega_{pe} \leqslant 4.4 \times 10^{12}$ s^{-1}. Therefore one may speak about the coincidence of the frequency of detected QEFs with the Langmuir-oscillation frequency.

Suppose nonetheless that the measured QEF frequency corresponds to some other (not Langmuir) oscillations. The highest-frequency oscillations are those at the frequency $\omega_{He} = eH/m_ec$. But even for them, assuming $\omega = \omega_{He}$ we find $H = 1.6 \times 10^7$ A/m. However, $I(t_I) \approx 200$ kA, $r \geqslant 1$ cm, so that in reality at the instant t_I we have $H \leqslant 3.2 \times 10^6$ A/m, which is almost one order of magnitude smaller than what the latter interpretation needs.

Finally the pressure decrease P_0/P_0' of 1.8 times leads to a 1.4-fold decrease in the QEF frequency ω/ω' Under the "Langmuir" treatment we have from this $N_e/N_e' \approx 2$, which corresponds well to the ratio $P_0/P_0' \approx 1.8$. For other oscillations (in particular, at the frequency ω_{He}) it is difficult to explain the decrease of QEF frequency since neither the current nor the radius of the plasma column have changed, and consequently the magnetic field has also not changed.

Thus the recorded QEF must be Langmuir oscillations. With the help of (4.2.46), the measured halfwidths of dips $\Delta\lambda_{1/2}^{dip}$ (nm) yield their amplitude E_0 (kV/cm):

$$E_0 = B[n_\nu^2 - (n_1 - n_2)_\nu^2 - m_\nu^2 - 1]^{-1}$$
$$\times |n_\alpha(n_1 - n_2)_\alpha - n_\beta(n_1 - n_2)_\beta|^{-1}(\Delta\lambda_{1/2}^{dip})_\nu, \tag{7.3.6}$$

where $B \approx 1760$ for D_α and $B \approx 3220$ for D_β. The measurements of $(\Delta\lambda_{1/2}^{dip})_\nu$ are most accurate at the most distant (from the line center) reference dips since they are broader than the near dips. Substituting the measured halfwidth of such dips at the D_α, D_β lines into (7.3.6) we find that $E_0(t_I) \approx 50\text{–}80$ kV/cm (the relatively large errors are caused by the insufficient spectral resolution). This is one order of magnitude larger than the thermodynamic equilibrium value of the amplitude of Langmuir oscillations in the plasma considered here.

In conclusion we would like to note that these diagnostic results have been used to explain the dynamics of a Z-pinch [7.16]. It was shown that, when one includes magnetic viscosity, even within single-fluid hydrodynamics rarefaction-type soliton solutions are possible. When electron inertia is taken into account, Langmuir oscillations of amplitude $E_0 \sim (m_e/M_i)^{1/2}H$ are developed at the front of the rarefaction-type soliton. As was pointed out in [7.16], this agrees qualitatively with the results of our investigations of the Z-pinch.

7.4 New Features of Intra-Stark Spectroscopy Caused by a High Density of Plasmas

Dips or depressions in line profiles have also been observed in experiments [7.17–19] where investigated hydrogen SL were emitted by dense plasmas: $N_e > 10^{18}$ cm^{-3}. At such densities, first, a relatively small increase of wave fields over the thermal level $E \approx [T/(e\pi r_D)]^{1/2}$ – an increase which would not be a surprise for pulsed discharges – could lead to quasistatic fields of order or greater than the mean ion microfield $\langle F \rangle \approx 8.8eN_e^{2/3}$. Second, a dipole approximation is no longer sufficient for describing the interaction of a radiator with perturbing ions: higher multipoles (at least, a quadrupole term) should be taken into account.

In [7.19] a theory of dips was further developed to allow for a high density of plasmas. It was shown that three qualitatively new features appear: 1) a significant shift of the dip positions (mostly to the blue side); 2) dips in the profile of central ("unshifted") Stark components; 3) multi-quantum-resonance dips. Then an experimental investigation of the L_α line of hydrogen was carried out in [7.19] employing a gas-liner pinch that operates as a modified z-pinch with a special gas inlet system. The implosion time was 3.0 μs, the decay time was 4.0 μs, the maximum electron density was 2.9×10^{18} cm^{-3}, and the maximum electron temperature was 12.5 eV. At each shot plasma parameters were measured by a coherent Thomson scattering.

An analysis of the entire set of data in [7.19] confirmed predictions of the theoretical part of [7.19] both qualitatively and quantitatively (Fig. 7.12). Experimental and theoretical blue shifts of the first- and the second-order dips were in a fairly good agreement. The shifts of the dip positions are much greater than the shifts of the center of gravity of the spectral line so that the former can be observed much more easily and reliably than the latter.

An experimental determination of distances between dips represents an easy and precise method for measuring a QEF frequency in dense plasmas. In particular, if the QEF corresponds to Langmuir waves, one may have a precise passive method for measurements of high electron densities (Fig. 7.13). In addition to

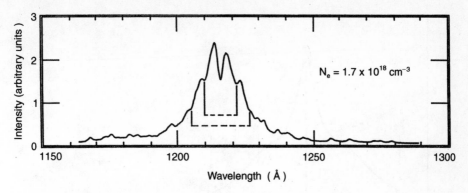

Fig. 7.12. Comparison of the experimental and theoretical positions of dips in the L_α line profile [7.19]. The theoretical positions of first and second order dips, respectively, are indicated by the pairs of vertical solid lines connected by a dashed line

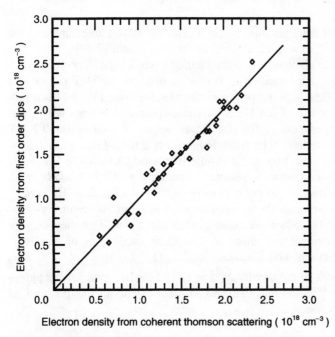

Fig. 7.13. Electron densities calculated from the separation of pairs of the first order dips versus the electron density measured by coherent Thomson scattering [7.19]

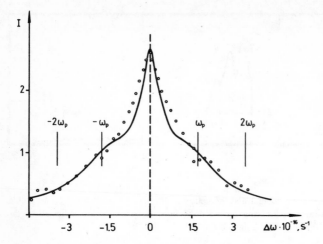

Fig. 7.14. The profile of the L_γ line of the Ne X ion in a laser-produced plasma: *points* – experimental data from [7.20]; *solid line* – theoretical calculations from [7.20]; *vertical lines* mark theoretically expected positions of depressions caused by Langmuir oscillations

that, an experimental determination of dip halfwidths allows an estimate of the amplitude of the QEF as well (2 MV/cm in the experiment [7.19]).

In conclusion, we emphasize that the method based on the intra-Stark spectroscopy (i.e. the dips-appearance effect) may be used for the QEF diagnostics even in more dense (and high-temperature) plasmas. So in the Fig. 7.14 is reproduced the L_γ line profile of ion Ne X in a laser-produced plasma which was registered in the experiment [7.20] (four laser beams of total power 0.2 TW in the impulse of duration 40 ps were focused on a neon-filled glass microsphere). By the smoothed form of the profile (the solid curve in the Fig. 7.14) using the calculated contours for plasmas containing no QEF *Yaakobi* et al. [7.20] have determined the electron density $N_e^{\text{width}} = (7 \pm 2) \times 10^{22} \text{ cm}^{-3}$. The analysis of structures in the original experimental profile (represented by points in the Fig. 7.14) which we have carried out, shows that they may be interpreted as the depressions caused by Langmuir oscillations of amplitude $E_0 = (550 \pm 80) \text{ MV/cm}$ and frequency $\omega_{\text{pe}} = (1.7 \pm 0.1) \times 10^{16} \text{ s}^{-1}$. This frequency corresponds to the density $N_e^{\text{dip}} = (10 \pm 1) \times 10^{22} \text{ cm}^{-3}$, what agrees with the value of N_e^{width} and thus confirms to some extent our interpretation of the experimental profiles.

7.5 QEFs in Tokamaks

7.5.1 Intense EFs in the Edge Plasma of the T-10 Tokamak

The peripheral plasma in tokamaks has recently been attracting considerable interest in fusion research because very slight changes in the parameters of this plasma can lead to a fundamental change in the structure of the overall plasma column and changes in the particle lifetime [7.21]. In this section we study the peripheral plasma in the T-10 tokamak through an analysis of the deuterium lines [7.22]. The deuterium emission spectrum was measured along a central chord in the equatorial plane of the tokamak. The diagnostic instrumentation, consisting of a ten-channel polychromator based on a MDR-2 monochromator with fiber optics and photomultipliers, makes it possible to measure the spectrum in a single discharge. The resolution of the individual channels is 0.6–0.7 Å. A polarizer is placed in front of the entrance slit of the polychromator for polarization analysis of the emission.

Figures 7.15, 16 show typical spectra of the deuterium lines D_α, D_β, and D_γ, in this case measured at a magnetic field of $B_0 = 1.65$ T. The most prominent feature in these profiles is a dip in the central part of the D_α and

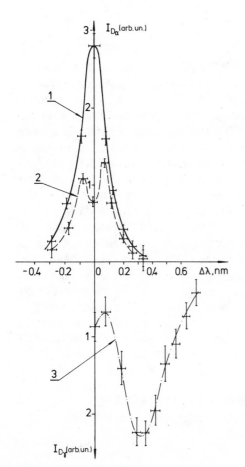

Fig. 7.15. Profiles of the D_α line *(curve 1 and 2)* and the red half of the profile of the D_γ line *(curve 3)*, measured at a discharge current $J = 180$ kA, at the tokamak magnetic field $B_0 = 1.65$ T, and at the electron density $N_e = 2.5 \times 10^{13}$ cm^{-3} [7.22]. *curve 1* – polarization of the emission is $e \perp B_0$; 2, 3 – $e \parallel B_0$

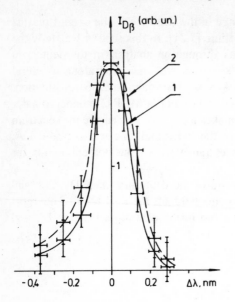

Fig. 7.16. Same as in Fig. 7.15, but for the D_β line. *curve 1* $- e \perp B_0, 2 - e \parallel B_0$

D_γ profiles, but not on the D_β profile.[3] The dips in the D_α and D_γ lines are observed on the profiles corresponding to emission with polarization parallel to the B_0 vector (π profiles). The dips in the deuterium polarization profiles cannot be explained either by a Zeeman splitting (since it would be smaller than the observed splitting by a factor of 2–2.5 and, furthermore, could not arise on π profiles) or by a self-absorption of the emission (since there are no dips on the D_α and D_γ profiles in the other polarization or in any of the D_β profiles). The only physical mechanism which might be responsible for the appearance of these dips is a Stark effect in anisotropic EFs. However, a problem remains to determine the type of these fields. The presence of dips in the π profiles of D_α and D_γ at $B_0 = 1.65$ T could in principle be explained by QEF with an amplitude $E_0 \gg m_e e \omega / n^2 \hbar$ (ω is the frequency of the fields and n is the principal quantum number) which are excited along the magnetic field of the tokamak, B_0 (Sect. 3.1). This interpretation, however, would require that a dip also be present in the π profile of D_β, which is at odds with the observations.

It turns out that the spectral features in the set of these polarization profiles can be explained on the basis of a superposition of a high-frequency QEF $E_0 \cos \omega t$, polarized along the observation direction, and a quasistatic EF $F \parallel B_0$

$$E(t) = e_z E_0 \cos \omega t + e_x F \qquad (7.5.1)$$

(e_x and e_z are unit vectors). In a situation with $\max(\omega, (n\hbar E_0 \omega / m_e e)^{1/2}) \gg n\hbar F/(m_e e)$ the imposition of field (7.5.1) on a hydrogen-like radiator gives rise

---

[3] In this section we use the word "dip" to simply mean "minimum" without any connection to the "resonance dips effect" discussed above.

to a fundamentally new spectroscopic effect described in Sect. 4.3: the QEF $e_z E_0 \cos \omega t$ *suppresses* the quasistatic EF $e_x F$ which is orthogonal to it. As a result, the effective quasistatic field acting on the level n of the deuterium atom is $e_x F_{\text{eff}}^{(n)} = F \cdot J_0(3n\hbar E_0/2m_e e\omega)$, $J_0(u)$ is the Bessel function. Furthemore, in the spectrum with z polarization, under the condition $3n\hbar E_0/2m_e e\omega \gtrsim 1$, satellites may exist at distance $\Delta\omega = \pm\omega, \pm 2\omega, \ldots$, from the center of the line; these satellites are comparable in intensity to the component at the frequency $\Delta\omega = 0$.

The presence of dips in the π profiles of D_α and D_γ and the absence of a dip from the π profile of D_β can thus be explained by the partial suppression, by the QEF $E_0 \cos \omega t$ of the quasistatic splitting of the $n = 2, 3, 5$ levels and *complete* suppression of the $n = 4$ level: $F_{\text{eff}}^{(4)} = 0$. Since when $F_{\text{eff}}^{(4)} = 0$ the most intense π and σ spectral components of the quasistatic profile of D_β are emitted at an unshifted frequency, $\Delta\omega = 0$, there is no dip at the center of the π profile of D_β. Under the assumption that for $n = 4$ the argument of the function $J_0(3n E_0/(2m_e e\omega)$ is the same as the first zero of this function, we find $E_0/(m_e e\omega) \approx 0.40$. Hence, under the condition that the frequency ω is near the electron cyclotron frequency ω_{Be} at $B_0 = 1.65$ T, we find $E_0 \approx 14$ kV/cm. In this case the values of $F_{\text{eff}}^{(n)}$ for the other levels are $F_{\text{eff}}^{(2)} \approx 0.67$ F, $F_{\text{eff}}^{(3)} \approx 0.34$ F, $F_{\text{eff}}^{(5)} \approx 0.26$ F. An estimate of the quasistatic fields from the π profile of D_α yields $F \approx 20$ kV/cm. It follows from Figs. 7.15, 16 that the intensities of the σ and π profiles integrated over the frequency differ by a factor of about two for the D_α line and are nearly the same for the D_β line. The probable explanation is that the populations of the Stark sublevels of the $n = 3$ level, which are split by the field $F_{\text{eff}}^{(3)}$, are not the same as their equilibrium values, while the Stark sublevels of the $n = 4$ level with $F_{\text{eff}}^{(4)} = 0$ remain equal in energy, and collisions cause their populations to become equal. The direction of the QEF $E_0 \cos \omega t$ is approximately along the observation direction, since the σ profiles of D_α and D_β have no intense satellites at the frequencies $\Delta\omega = \pm\omega, \pm 2\omega, \ldots$ A determination of the ion temperature from the measured halfwidth of the spectral component of the profile of D_α yields $T_i \sim 10$–15 eV, in agreement with the estimates $T_i \sim 10$–30 eV found for the region of the maximum intensity of the deuterium lines by other methods, for various operating conditions in the T-10.

At a stronger magnetic field, $B_0 = 3.05$ T, the profiles of the deuterium lines change substantially: it is the σ profile of D_α, not the π profile, which has the greater broadening and the central dip (Fig. 7.17). These results mean that the quasistatic field is now directed along the y axis (perpendicular to B_0 and to the direction of observation). A quantitative analysis of the σ profile of D_α in Fig. 7.17 yields $F \sim 20$ kV/cm (and, as before, its frequency $\Omega \ll \omega_{\text{Be}}$), $0 \leqslant E_0 \lesssim 10$ kV/cm.

Thus an analysis of the experimental data on the basis of this spectroscopic effect shows that oscillations grow to EFs of 10–20 kV/cm, at least in the peripheral plasma of the T-10 tokamak under conditions of high current and comparatively low density. We note in conclusion that the quasistatic component

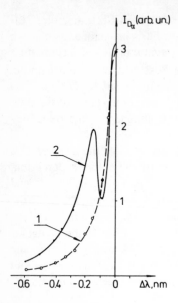

Fig. 7.17. Profiles of the D_α line (blue half) measured at $J = 450$ kA, $B_0 = 3.05$ T, and $N_e = 6 \times 10^{13}$ cm^{-3}. *curve 1 – e* $\parallel$ B_0; 2 – profile measured without a polarizer. From [7.22]

does not necessarily represent a regular, one-dimensional, low-frequency quasi-monochromatic EF $F \cos(\Omega t + \varphi)$. The results of our analysis of the experimental profiles of D_α, D_β, and D_γ are also consistent with the conclusion that the quasistatic component represents a low-frequency turbulence with a quasi-two-dimensional spectrum, which has developed in the xy plane ($B_0 = 1.65$ T) or the yz plane (at $B_0 = 3.05$ T).

7.5.2 Novel Spectroscopic Diagnostics of EFs in Tokamaks

We begin by analyzing the possibilities of corpuscular-spectroscopic diagnostics of QEFs $E(t) = E_0 \cos \omega t$ in tokamaks. These fields may be microwaves from the external source used for additional heating or some regular waves excited in a plasma (as was experimentally revealed in the T-10 tokamak, Sect. 7.5.1). But for the conditions in which a beam of neutral atoms with velocity v is injected across a magnetic field B the beam atoms experience a strong Lorentz field $F = v \times B/c$ of order 10–100 kV/cm. It was already experimentally demonstrated that this strong static field F leads to a significant shift [7.23] and/or splitting [7.24] of SLs. Thus the question arises whether it is possible to detect relatively weak QEFs $|E(t)| \ll F$ in spite of a strong field F by analysis of spectra of beam atoms. Our main idea is to use some spectroscopic manifestations which may be caused only by *joint action* of a static field F and a dynamic field $E(t)$.

The first proposal is to utilize a lithium beam and then to observe the profile of the SL of Li I at 4603 Å (2P - 4P, D, F). As a starting point, recall the results of Sect. 6.2.1 on Baranger–Mozer satellites of dipole-forbidden SLs caused by the allowance for an additional static field F. It was shown in Sect. 6.2.1 that even relatively weak fields $F \lesssim 10^{-1}E_0$ drastically change the spectra; in particular, the field F leads to the appearance of intense satellites separated by $\pm\omega$ from a dipole-allowed SL (this type of satellite was formerly known only in spectra of hydrogen SLs under the action of QEFs).

In our present case the strong field F completely mixes the states 4F, 4D, 4P of the Li atom and they respond to a weak field $E(t)$ as a hydrogen-like structure displaying pronounced satellites separated by $\pm\omega$ from the line components 2P–4D and 2P–4F. A numerical example: the Li-beam with an energy of 100 keV is injected across the magnetic field $B = 2.8$ T (so that $F = 46$ kV/cm). If $E(t)$ represents the oscillations at the electron cyclotron frequency, then the satellites appear separated by $\pm\lambda_\omega = \pm\omega\lambda_0^2/(2\pi c)$ from the position λ_0 of the main components. Their relative (with respect to the main component) intensity is $S/I_a \approx [0.2E_0 \text{ [kV/cm]}]^2$. Thus when $E_0 \gtrsim 1.5$ kV/cm we have $S/I_a \gtrsim 0.1$, i.e. an effect which should be easily observable. In order for the Doppler width of all features (limited by the beam divergence) to be smaller than λ_ω, the angular divergence of the beam should be $v_\perp/v_\parallel < 10^{-2}$.

The second proposal is to use a hydrogen beam and then to observe the profile of the SL of H_β at 4861 Å. A strong field F splits the degenerate hydrogen states which can then interact with a weak field $E(t)$, leading to the appearance of pronounced forbidden components (unresolved satellites) in the line center. Recall that under the action of a static field only ($E_0 = 0$) the SL H_β has no central Stark components. A numerical example: an H-beam of 40 keV energy is injected across a magnetic field $B = 1.4$ T (so that $F = 40$ kV/cm). The separation between Stark components of the H_β is $\Delta\lambda_F = 0.030\,F\text{[kV/cm]} = 1.2$ Å. The relative increase of the intensity in the center of H_β (with respect to the main maxima) after switching on a microwave field $E(t)$ is $I_f/I_a \sim (6E_0/F)^2$. Thus when $E_0 \gtrsim 2$ kV/cm we have $I_f/I_a \gtrsim 0.1$, i.e., a rather obvious effect. In order for the Doppler broadening to be smaller than the Stark one, the angular divergence of the beam should be $v_\perp/v_\parallel < 2.2 \times 10^{-2}B(T) \approx 0.03$.

In both these proposals it was assumed that the splitting of a multiplet in a Lorentz field dominates over pure Zeeman splitting. This is the case when the beam velocity v is greater than $c/205.5n$. For the principal quantum number $n = 4$ the corresponding condition for the beam energy is $E_\text{beam} > E_\text{cr}$, where $E_\text{cr} = 4.9$ keV for Li and $E_\text{cr} = 0.7$ keV for H.

A new spectroscopic method for local determination of the effective charge Z_eff in tokamaks has been theoretically proposed [7.25] and recently experimentally tested [7.26]. The method is based on two ideas. The first is that under tokamak conditions the broadening of hydrogen SLs by protons and by impurity ions is not quasistatic but impact broadening, the impact width Γ being proportional to Z_eff. But Γ is approximately one order of magnitude smaller than the Doppler width $\Delta\omega_{1/2}^D$. The second idea was to use laser saturation of an excited transition (with a laser line width $\Delta\omega_{1/2}^r \ll \Delta\omega_{1/2}^D$) in order to extract the small value of Γ in spite of the large value of $\Delta\omega_{1/2}^D$. As the intensity of the laser light E_0 increases, an increase of the fluorescence SL width (so-called power broadening): $\Gamma_B = \Gamma(1 + G)^{1/2}$, $G \equiv (d_{12}E_0/\hbar)^2/\gamma\Gamma$, occurs where γ is the width of the emitting. This means that Doppler and power broadenings may be distinguished from the profiles of a hydrogen fluorescence line, so that the values of Γ and Z_{eff} can be measured. An alternative experimental technique is based on the Doppler-free two-photon excited fluorescence. A high quality laser spectrometer allowing to realize this diagnostics with an acceptable signal-to-noise ratio has been recently developed and tested on a simulation plasma device [7.50].

Recently we have developed a detailed, consistent theory of ion impact broadening [7.51] that has two distinctive features. First, we took into account a strong magnetic field $B \gtrsim 1T$ characteristic for tokamaks. Second, our theory is free from a shortcoming of the standard semiclassical theories of Stark broadening which were intrinsically divergent at small impact parameters [7.25, 52]. As a result we have shown that for the laser-induced fluorescence measurements of the effective charge $Z_{eff} \equiv \sum Z_i^2 N_i/N_e$, the best choice is the central (π) component of Zeeman triplet of the SL L_α.

Our third proposal is to further improve this method by carrying out analogous laser saturation spectroscopy measurements but on hydrogen atoms in a beam injected into the plasma. The advantages are the following: 1) the uncertainties intrinsic to the method [7.26] vanish, since the proportionality coefficient between Γ and Z_{eff} is now controlled by the known beam velocity v_{beam} (instead of a mean hydrogen atom velocity $\langle v_a \rangle$, which usually is not well-known experimentally); 2) the Doppler width is drastically reduced (since it is now determined by the beam divergence $v_\perp/v_\parallel \ll 1$), allowing laser saturation of isolated Stark components of a SL. The enhancement of component widths by the laser field will become more pronounced.

The formula for Γ in this case significantly differs from that in [7.25,26] and may be expressed as follows:

$$\Gamma = \{12(\hbar/m_e)^2 f(n, n') N_e v_{beam}^{-1} \ln[2em_e^2 c v_{beam}/3n^3\hbar^2 B)]\} Z_{eff},$$

$$f(n, n') \equiv n^2[n^2 + (n_1 - n_2)^2 - m^2 - 1] + n'^2[n'^2 + (n_1' - n_2')^2 - m'^2 - 1]$$
$$- 4nn'(n_1 - n_2)(n_1' - n_2'), \tag{7.5.2}$$

where n_1, n_2, m and n_1', n_2', m' are parabolic quantum numbers of upper and lower levels correspondingly; N_e is the electron density.

Our fourth proposal is to combine the saturation technique not with H – but with He- or Li-beams. The suitable SLs are He I at 4471 Å and 4922 Å or Li I at 4603 Å (all transitions are 2P–4D). The general expression for Γ is

$$\Gamma = \{[8\pi N_e/(3\langle v_{rel} \rangle)][\hbar r_{12}/(m_e a_0)]^2$$
$$\times \ln[\max(1, m_e a_0\langle v_{rel} \rangle^2/\hbar r_{12}\Delta\omega_{12})]\} Z_{eff}, \tag{7.5.3}$$

where $\Delta\omega_{12}$ and r_{12} are the dipole matrix element and the separation between the states 4D, 4F. For thermal He- or Li-atoms the relative perturber-radiator velocity $\langle v_{\text{rel}} \rangle \approx v_{\text{Ta}}$, so that the logarithm in (7.5.3) is equal to zero (no impact broadening). But for He- or Li-atoms in a beam, $\langle v_{\text{rel}} \rangle \approx v_{\text{beam}}$, so that the ratio $m_e a_0 \cdot \langle v_{\text{rel}} \rangle^2 / \hbar r_{12} \Delta \omega_{12}$ increases by $E_{\text{beam}} / T_a \sim 10^2 - 10^3$ times, which leads to a significant value of Γ. In other words, the beam may change the mechanism of He- or Li-line broadening by impurity ions from quasistatic to impact. Due to the strong Lorentz field the allowed and forbidden components have comparable intensities. Hence laser saturation measurements of Z_{eff} may be carried out on both 2P–4D and 2P–4F components.

7.6 QEFs in Plasmas Interacting with a Strong Microwave Field

7.6.1 Technique Utilizing Hydrogen or Deuterium Lines

In experiments described in [7.27] a target deuterium plasma was created in a quartz tube placed in an S-band waveguide. The plasma diffused along the lines of a superimposed axial magnetic field to the point of observation. The magnetic field was chosen to be 0.145 T, well above the electron cyclotron resonance with respect to the plasma-generating magnetron ($\nu = 2.45$ GHz). The target plasma was pulsed at 300 Hz at a pulse width of 500 μs. The pressure was in the range of 0.1–1 Pa, the averaged electron density was $N_e \leqslant 5 \times 10^{12}$ cm^{-3}, and the electron temperature in the range of 3–9 eV.

In the observed region of the discharge tube, 34.8 GHz microwaves were focused into the plasma with a horn-lens system. Optical measurements were performed side-on. The profiles of the deuterium lines D_β, D_γ, D_δ under the influence of the microwave QEF were compared with profiles measured immediately after the microwave pulse had passed (Fig. 7.18). Thus it was possible to obtain the satellite profiles by subtracting the undisturbed line intensities from the ones taken while the field was turned on. The distance of the microwave satellites from the line center was seen to be a multiple of the microwave field frequency. The satellite structure turned out to be solely due to the action of QEFs, without the disturbing influence of static fields ("pure" Blochinzew-type satellites). So the satellite pattern remained simple and the satellite intensity allowed determination of the QEF amplitude.

In addition, measurements of D_β with a polarizer were made, showing a great sensitivity of the satellite intensity to the polarization direction (which could, in principle, be used to determine an unknown polarization direction). All these measurements of first and second order satellites and for different polarizations have led to consistent values of the QEF amplitude: $E_0 = 1.00 - 1.35$ kV/cm.

Intracavity laser (ICL) spectroscopy is a highly sensitive technique which can record and accurately measure the intensity of the spectral components in the presence of QEFs. The experiments described in [7.28] were carried out in a pulsed discharge plasma in a D–H mixture at a total pressure of

I, counts ×10³

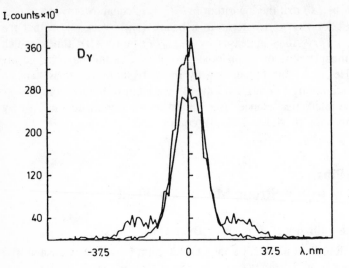

Fig. 7.18. The line profile of D_γ with a microwave QEF turned on (1) and turned off (2) [7.27]

70–1300 Pa. The plasma was generated in sealed-off U-shaped glass tubes of 8 mm diameter by a current pulse generator (pulse length 0.1–5 ms, amplitude 10^{-2}–10 Å). The intense microwaves were generated by a cyclotron resonance maser (gyroton) with an output power of 200 kW per pulse and an operating frequency of 38.5 GHz (the pulse was 200 μs long). The gas discharge tube was equipped with optical windows at each end and was placed inside the laser cavity, which was bounded by mirrors. Flashlamps pumped a cell filled with a solution of oxazine-17 dye (for spectral measurements near the H_α line) or coumarin-47 (for the H_β line) in ethanol; this enabled one to generate pulses of ≈ 1.5 μs and spectral width ≈ 8 nm.

It was shown in [7.29] that the experimentally measurable ratio $I(\nu)/I_0$ of the intensities inside and outside the absorption line averaged over τ is determined by the formula

$$I(\nu)/I_0 = \{1 - \exp[-\gamma(\nu)]\}/\gamma(\nu), \quad \gamma(\nu) = \kappa(\nu)lc\tau/L. \quad (7.6.1)$$

Here κ is the absorption coefficient, l is the length of the cell, τ is the laser pulse duration, and L is the length of the cavity. Like any technique based on absorption, ICL spectroscopy has a limited dynamic range because the measurement errors increase for both small ($\gamma < 0.05$) and large ($\gamma > 3$) values of the effective optical thickness. One can optimize the conditions for measuring the weak spectral components by selecting the laser pulse and the concentration of the absorbing atoms in the discharge [this concentration determines the value of κ in (7.6.1)].

In most cases (except for very strong EF), near the center of the unperturbed line, κ exceeds the gain of the dye, so that lasing stops at these frequencies. In

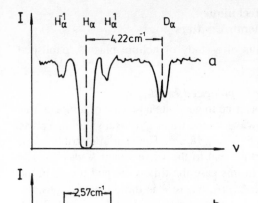

Fig. 7.19. The profiles of the H_α and D_α lines and their satellites recorded by intracavity laser spectroscopy. In **b** the absorbing atoms density is larger than in **a**. From [7.28]

the experiments therefore the plasma was generated in an H–D mixture with a partial pressure ratio $P_{H_2}/P_{D_2} = 5$–50. This ratio was chosen so that the intensity of the satellites of the H_α line in the microwave field was roughly equal to the intensity of the D_α line. By using an H–D mixture the authors of [7.28] were able to determine the frequency scale of the recording system.

Figure 7.19 shows some typical ICL spectrograms near the α-lines of the Balmer series for two concentrations of the absorbing atoms. The absorption coefficients for the D_α line and $H_\alpha^{\pm 1}$ satellite are comparable in order of magnitude (Fig. 7.19a). Since the partial pressure ratio P_{H_2}/P_{D_2} is known, in this case one can quite accurately determine the ratio of the absorption coefficients for the H_α and $H_\alpha^{\pm 1}$ spectral components. Similar measurements can be used to find the strengths of QEFs in the plasma. According to (7.6.1), increasing the density of absorbing atoms (or the laser pulse length τ) will enhance the dip in the laser spectrum. Although the $D_\alpha^{\pm 1}$ satellites are also observed (Fig. 7.19b), the error in measuring the absorption coefficient for D_α is very large.

For laser and microwave fields polarized in the same direction, the satellites for the unperturbed lines were considerably more intense than for crossed laser and microwave fields. This can be exploited for polarization measurements, which are generally highly accurate and sensitive.

The authors of [7.28] made no attempt to measure the absolute values of the absorption coefficients. Results such as the ones shown in Fig. 7.19 therefore yield only relative values of $\gamma(\nu)$.

There are, of course, other experiments in which SLs of Coulomb radiators were used for nonlocal diagnostics of powerful microwaves in plasmas. Due to lack of space, however, we only mention some additional references [7.30–32].

7.6.2 Quasilocal Measurements Technique
Utilizing Lines of Non-Coulomb Emitters

In this section we present the results of a study of helium plasma, published in [7.33].

Observations of Helium SLs, Using Unpolarized Light
An inductive discharge in helium was used to generate a plasma of temperatures $T_e \approx 10^5$ K, $T_i \approx 10^4$ K. The microwave source was a gyrotron yielding a pulse of 200 μs duration at the wavelength $\lambda = 0.78$ cm. The spatial distribution of the QEF amplitudes in the plane orthogonal to the wave vector $\boldsymbol{k}$ was $E(r) = E_0 \exp(-r^2/a^2)$, where $a \approx 1$ cm. In this case the time-averaged power in the beam focus is equal to $P = \varepsilon^{1/2} c a^2 E_0^2/16$, where ε is the dielectric constant. For $\varepsilon = 1$ we can use the convenient expression relating E_0 [V/cm] and P[W],

$$E_0 \approx 22 P^{1/2}. \tag{7.6.2}$$

During the microwave pulse the plasma density gradually increased and reached the critical value (for the wave number $k = 1.28$ cm^{-1}) $N_{cr} \equiv m_e \omega/4\pi e^2 = 1.85 \times 10^{13}$ cm^{-3}. The spectra near the SLs of He I at 4922 Å and 4471 Å, exhibiting pronounced satellites, were recorded at some time t_1 when the plasma is transparent ($N_e < N_{cr}$ i.e., $\varepsilon \approx 1 - N_e/N_{cr} > 0$), and at t_2 when $N_e = N_{cr}$ (i.e., $\varepsilon = 0$) is reached and electromagnetic energy is nonlineraly absorbed.

A monochromator collected the satellite emission from a relatively small volume v_E (where microwaves were present) with the linear dimension $l_E \approx$ 1 cm and the allowed SL emission was detected from a nonperturbed plasma volume $V \gg V_E$ with a linear dimension $L \gg l_E$. Therefore the measurements of the ratio of near and far satellite intensities S_-/S_+ allowed in fact, a determination of the local QEF amplitude E_0 in the region V_E using the results of Sect. 5.1.2.

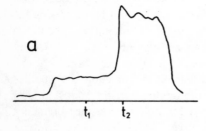

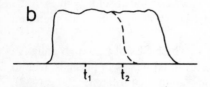

Fig. 7.20. (a) Typical emission signal of the near satellite of the He I 4922 Å line. (b) Incident (*solid curve*) and transmitted (*dashed*) microwave signal. The time t_2 corresponds to $\varepsilon = 0$ ($N_e = N_{cr}$). From [7.33]

Figure 7.20 shows a typical signal of the emission of the near satellite of the He I 4922 Å line. The steep increase of the satellite luminescence at t_2 (while the emission of the allowed line remains practically constant) indicates an increase of the QEF amplitude in the plasma.

The optical measurements were made at pump microwave powers $P = 1.9$, 3.8, 7.5, 15, and 60 kW. Figures 7.21, 22 show emission spectra recorded,

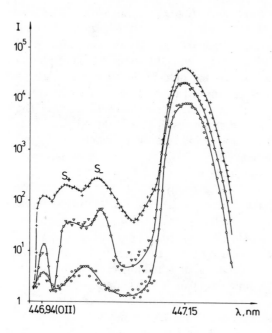

Fig. 7.21. Spectrum near the He I 4471 Å line at a pump microwave power of $P = 60$ kW for three plasma regions; (+) – region of plasma resonance ($N_e = N_{cr}$, $\varepsilon = 0$); (∇) – plasma transparency region ($N_e < N_{cr}$, $\varepsilon > 0$); (0) – region without microwaves. The O II 4469 Å line can also be seen. From [7.33]

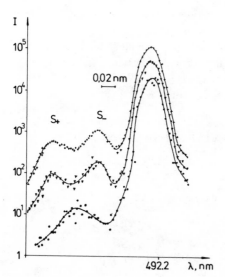

Fig. 7.22. The same as in Fig. 7.21, but near the He I 4922 Å line

respectively, near the lines of He I 4471 Å and 4922 Å at the times t_1 and t_2 at a power $P = 60$ kW.

We note that near the 4471 Å line not only two satellites are observed, but also the O II 4469 Å line (owing to oxygen occlusion). Note that at low powers, $P \ll 10$ kW, the satellite emission is weak and in practice one observes near the forbidden line only the O II 4469 Å line, which can be erroneously interpreted as a "solitary" satellite. This is possibly the case in [7.34].

The spectra near the 4471 Å and 4922 Å lines, recorded at the time t_2 and at powers $P = 7.5$, 15, and 60 kW, were analyzed by measuring the ratio S_-/S_+. A subsequent comparison with the theoretical field dependences of S_-/S_+ (Fig. 5.4) yielded the amplitude E_{op} of the QEF in the region of plasma resonance ($\varepsilon = 0$). The amplitude E_{or} of the field in the transparent plasma (at the instant t_1, when $\varepsilon > 0$) was determined from the known pump power P with the aid of (7.6.2) rather than from the experimental ratio S_-/S_+. The reason is that in the transparent plasma at powers $\lesssim 15$ kW the far satellite is weak and its intensity is subject to large errors (the same situation is typical of the plasma resonance region at powers $\lesssim 4$ kW). At $P = 60$ kW, however, measurement of S_-/S_+ in a transparent plasma yielded $E_{or} = 4.6 \pm 0.4$ and 6 ± 1 kV/cm for the lines 4922 and 4471 Å, respectively, which agrees with the value $E_{or} = 5.4 \pm 0.5$ kV/cm obtained from (7.6.2). The results, including the field gain $K = E_{op}/E_{or}$, are given in Table 7.1.

We note that in the case of the He I 4471 Å line the emission of the O II 4469 Å impurity overestimates the intensity S_+ of the far satellite, and hence the amplitude E_0. Therefore in the calculation of K the higher reliability of the data obtained for the He I 4922 Å line was taken into account.

The QEF amplitudes were also estimated from the measured ratio S_-/I_a in order to illustrate the substantial effect of the spatial factor, as well as to estimate roughly the field at low powers $P < 4$ kW, at which the ratio S_-/S_+ cannot be measured reliably. The results averaged over two lines (4922 Å and 4471 Å) are shown in Fig. 7.23.

The straight line in Fig. 7.23 corresponds approximately to the case of a transparent plasma. It was drawn through the vertical error bar of E_{or} determined from the ratio S_-/I_a measured at the time t_1 for $P = 60$ kW. It follows from the slope of this line that the real value of the amplitude E_0^{real} is approximately 2.8

Table 7.1. Experimental conditions and resulting local QEF amplitudes in a helium plasma determined from two helium emission lines. From [7.33]

P [kW]	E_{or} [kV/cm]	E_{op} [kV/cm]		$K = E_{op}/E_{or}$
		4922 Å	4471 Å	
7.5	1.9 ± 0.2	6.4 ± 0.4	$6.5^{+0.5}_{-1}$	3.4 ± 0.6
15.0	2.7 ± 0.3	6.1 ± 0.4	$7.0^{+0.5}_{-1}$	2.4 ± 0.5
60.0	5.4 ± 0.5	6.2 ± 0.4	$8.0^{+0.5}_{-1}$	1.3 ± 0.3

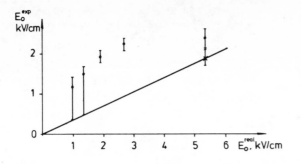

Fig. 7.23. Relationship between a field E_0^{exp} determined experimentally by measuring the ratio S_-/I_a for the lines 4922 Å and 4471 Å of He I and a real field E_0^{real}; this relationship results from the presence of a spatial form factor. From [7.33]

times larger than E_0^{exp} determined by measuring S_-/I_a. We can then estimate roughly the spatial form factor at $L/l_E \sim (2.8)^2 \approx 8$. The obtained value $L \sim 8l_E \sim 8$ cm agrees with the halfwidth of the spatial distribution of the plasma density over the chamber diameter.

The error bars above the line in Fig. 7.23 correspond to the plasma resonance ($\varepsilon = 0$) at powers $P = 1.9, 3.8, 7.5, 15,$ and 60 kW. For a rough estimate of the real amplitude E_0^{real}, each value of E_0^{exp} should also be multiplied by 2.8. At $P = 7.5, 15,$ and 60 kW, the estimates of E_0 obtained in this manner can be compared with the more rigorously derived results in Table 7.1 and the agreement between them verified.

The results offer evidence that the EF gain in the plasma resonance region ($\varepsilon = 0$) has a maximum at $E_{0r} \sim 2$ kV/cm and decreases with increasing field in the pump wave. The amplitude of the EF E_{0p} does not exceed $E_{\max} \sim 7$ kV/cm. It is of interest to determine not only the amplitude but also the predominant direction of the EF at $\varepsilon = 0$. This is made possible by the polarization technique presented below.

Observations of the Polarization of Helium Lines
To determine the predominant direction of the EF we placed a polarizer in front of the monochromator. The radiation intensity of the near satellite of the He I 4922 Å line was recorded at two positions of the polarizer transmission axis, along the EF in the pump wave (this coincides with the direction of the quasi-stationary magnetic field in the plasma) $S_-^{(z)}$, and perpendicular to the direction (i.e. along the wave vector $k \perp B$) $S_-^{(x)}$. Thirty data points were averaged for each measurement of the polarized radiation intensity $S_-^{(z)}$ or $S_-^{(x)}$ The results (including the standard deviation) are given in Table 7.2.

Table 7.2. Incident microwave powers and resulting polarization ratios of the 4922 Å He I emission line satellite.

P [kW]	1.3	4	13	40
$S_-^{(z)}/S_-^{(x)}$	1.008 ± 0.026	0.956 ± 0.014	0.942 ± 0.017	0.994 ± 0.014

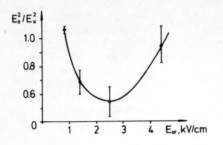

Fig. 7.24. Experimental dependence of the ratio E_z^2/E_x^2 in the region of the plasma resonance on the pump field E_{0r}. From [7.33]

Using these data and (5.1.22) we determined the value of $\cot^2\gamma \equiv E_z^2/E_x^2$. This procedure first required the determination of the amplitude E_{0p} (by the procedure described above), determination of $f_-(E_{0p})$ (Fig. 5.5) and only then followed the direct calculation of $\cot^2\gamma$ from (5.1.22). The obtained dependence of E_z^2/E_x^2 on the pump field E_0 is shown in Fig. 7.24. It can be seen that the enhancement of the EF in the plasma-resonance region is due to a considerable degree to the appearance of the E_k component (parallel to k), which was practically nonexistent in the pump wave. This effect is the largest at $E_{0r} = 2.5$ kV/cm. With increasing pump field, the longitudinal and transverse components of the field E_{0p} become of the same order ($E_z \approx E_x$).

We also determined the predominant direction of the pump field in a transparent plasma at $P = 40$ kW. Statistical processing of the data yielded $S_-^{(z)}/S_-^{(x)} = 1.183 \pm 0.026$. The measurements were made at a plasma density $N_e \approx 0.5 N_{cr}$, corresponding to $\varepsilon \approx 0.5$. In this case we have in lieu of (7.6.2) E_0 [V/cm] $\approx 26 \, (P \, [\text{W}])^{1/2}$. A power $P = 40$ kW corresponds to an amplitude $E_0 \approx 5.2$ kV/cm, at which $f_-(E_0) \approx 1.24$. Using (5.1.22), we obtain $E_z^2/E_x^2 = 8_{-3}^{+8}$. The relatively large spread is due to the fact that according to (5.1.22) $\cot^2\gamma \propto [f_-(E_0) - S_-^{(z)}/S_-^{(x)}]^{-1}$, and the values of f_- and $S_-^{(z)}/S_-^{(x)}$ are then close.

It was known beforehand that the E_Z component is the largest in the focused microwave beam, and that in the $y = 0$ plane

$$E_z = E_0 \exp(-z^2/a^2) \exp(iwt - ik_0\varepsilon^{1/2}x). \tag{7.6.3}$$

Such a beam, however, should also have an E_x component. In fact, substituting (7.6.3) into the condition $\operatorname{div} E = 0$ we obtain

$$E_x = (2zE_0/k_0\varepsilon^{1/2}a^2) \exp(-z^2/a^2) \exp(iwt - ik_0\varepsilon^{1/2}x + i\pi/2). \tag{7.6.4}$$

Consequently $|E_x^2/E_z^2| = (2z/k_0\varepsilon^{1/2}a^2)^2$. Calculating the mean spatial value we get $\langle|E_x^2/E_z^2|\rangle \approx 1/16$, which agrees with the result of polarization measurements within the standard error.

We note in conclusion that the present polarization analysis results, when the function $f_-(E_0)$ in (5.1.22) is included, differ substantially from the results that would be obtained from the corresponding formula of [7.35] (i.e., replacing f_- by 4/3). In particular, the minimum of the curve in Fig. 7.24 (which demonstrates the effect of field enhancement along k) would be reached at $E_{0r} = 1.4$ kV/cm instead of $E_{0r} = 2.5$ kV/cm, and the result of the control experiment for a microwave beam in a transparent plasma would be $E_z^2/E_x^2 = 4 \pm 1$, which is very different from the true value $\langle |E_x^2/E_z^2| \rangle \approx 1/16$.

Summary

The results of analysis of the spectra of several helium lines, both unpolarized and polarized, permit the following conclusions to be drawn.

Amplification of the QEF, $E_{0p} > E_{0r}$, takes place when the threshold field $E_{or}^{min} \sim 0.5$ kV/cm is exceeded in the pump wave near the critical density surface ($N \approx N_{cr}, \varepsilon = 0$). The function $E_{0p}(E_{0r})$ is nonlinear. The amplitude gain $K = E_{0p}/E_{0r}$ has a maximum at $E_{0r} \sim 2$ kV/cm. When the pump field is increased the gain "saturates" and the amplitude E_{0p} does not exceed $E_{max} \sim 7$ kV/cm. For a plasma with $N_e \sim 2 \times 10^{13}$, cm^{-3} and $T_e \sim 10$ eV ($T_i \sim 1$ eV) this means that the energy density of the intraplasmic QEF is $W \lesssim 10^{-1} N_e T_e$. These conclusions agree with results of spectroscopic investigations at the same installation based on an analysis of the profiles of the hydrogen line H$_\beta$ [7.31].

The polarization measurements have shown that in the region of plasma resonance there is a preferred amplification of the longitudinal component of the EF (parallel to the vector k of the pump wave). This effect is the largest at a pump amplitude of $E_{0r} \approx 2.5$ kV/cm, at which the ratio of the energy densities of the longitudinal field components for $\varepsilon = 0$ and for $\varepsilon > 0$ is $\sim 10^2$.

The earlier experiments with this setup [7.36] revealed a number of other nonlinear effects: generation of fast electrons, development of ion-acoustic oscillations, and superthermal electromagnetic radiation near the plasma frequency. All these facts allow us to conclude that the action of a powerful electromagnetic wave on a collisionless plasma produces a parametric instability and excites low-frequency plasma turbulence and intense Langmuir oscillations whose energy density is much higher than in the pump wave.

As for other experiments in which passive diagnostics based on observations of helium SL satellites were used to study a plasma interacting with powerful microwaves, we would like to call attention to [7.37].

7.6.3 Techniques of Local Laser Fluorescence Diagnostics

Hydrogen Lines

The active diagnostics method described in Sect. 3.4 was used in the experiments reported in [7.38]. In this work the same microwave source was used as in the investigations discussed in Sect. 7.6.2. At the focus of the microwave beam a discharge tube (filled with hydrogen) was installed in which a DC discharge was ignited. A dye laser was tuned to the wavelength of the H$_\alpha$ line (6563 Å). The laser power was 20 kW, the pulse duration 20 ns, and the bandwidth of the laser was 0.08 Å. The EF vectors of the laser (E_1) and microwave (E) radiation were parallel, the wave vectors k_1 and k were orthogonal. The spectroscopic data were obtained at a hydrogen pressure $P_0 = 670$ Pa, an electron concentration $N_e = 6 \times 10^{11}$ cm^{-3} and a temperature $T_e = 2 \times 10^4$ K, and gas temperature $T_a = 400 \pm 40$ K.

Figure 7.25 shows the dependence of the inverse value of the fluorescence signal B^{-1} on the inverse intensity of the laser radiation in the absence and presence of microwaves. It is easy to find the experimental ratio of the slopes, $g_0(\varepsilon) = \tan\alpha(P = 0)/\tan\alpha(P)$. Further, by using the predicted theoretical function $g_0(\varepsilon)$ for the H$_\alpha$ line (Fig. 3.11) one can determine the value of $\varepsilon = 3\hbar E_0/2m_e e\omega$ (ω is the microwave frequency) and the microwave amplitude E_0 for each value of P. The results are presented in Table 7.3.

In parentheses in Table 7.3 are the E_0 values calculated by (7.6.2), which relates the values of E_0 and P in vacuum. The good agreement between the measured and calculated amplitudes confirms the effectiveness of the presented method. Moreover the agreement means that under these experimental conditions the transverse relaxation time T (and consequently the impact

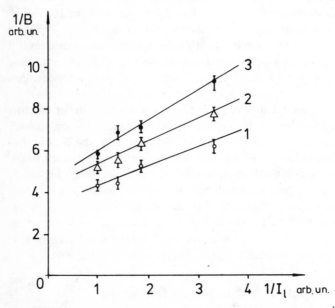

Fig. 7.25. Dependence of the inverse value of the fluorescence signal B^{-1} on inverse intensity of the laser radiation I_l^{-1} at different microwave powers P: *curve 1*: $P = 0$, *curve 2*: $P = 45$ kW, *curve 3*: $P = 60$ kW. From [7.38]

Table 7.3. Microwave amplitudes measured in a hydrogen plasma for different incident powers

P [kW]	45	60
E_0 [kV/cm]	4.1 ± 0.7	6.0 ± 0.9
	(4.7 ± 0.2)	(5.4 ± 0.2)

halfwidths of levels $n = 2$ and $n = 3$) do not depend on the microwave amplitude E_0.

Let us consider the applicability of the method to the diagnostics of microwaves of lower frequency $\omega/2\pi c \ll 1$ cm^{-1}. According to Sect. 3.4 the satellite halfwidth is $\Delta\omega_{1/2}^{(r)} = 2(1 + G_r)^{1/2}/T$. At the first glance the method cannot be used when $\omega < \Delta\omega_{1/2}^{(r)}$ (in the experiment of [7.38], $\omega/2\pi c = 1.3$ cm^{-1}, $\Delta\omega_{1/2}^{(0)}/2\pi c = 0.5$ cm^{-1}). However, it should be noted that the Bessel function argument in (3.4.5) depends on E_0/ω and when its value is much greater than unity, the most intense satellites are of the index (Sect. 3.1)

$$r_{max} \approx A\varepsilon - 0.81(A\varepsilon)^{1/3}, \quad \varepsilon = 3\hbar E_0/2m_e e\omega, \tag{7.6.5}$$

where $A = 3$ for H_α and $A = 8$ for H_β. Therefore at large values of E_0/ω the laser should be tuned to the frequency $\pm r_{max}\omega + \omega_{ab}$. Then the applicability of the method is controlled by the inequality $r_{max}\omega > \Delta\omega_{1/2}^{(r)}$ or, allowing for the Doppler effect, by the more general inequality

$$r_{max}\omega > \max(\Delta\omega_{1/2}^{(r)}, \Delta\omega_{1/2D}), \tag{7.6.6}$$

where $\Delta\omega_{1/2D}$ is the Doppler halfwidth of the SL. It is seen that the higher the microwave amplitude, the better one can achieve the conditions of (7.6.6).[4]

Note that the developed diagnostics method can be used in low-temperature plasmas for local measurements of Langmuir wave fields and other fields as well (and not only for microwaves penetrating into a plasma from an external source). In high-temperature plasmas, on the basis of Sects. 3.4, and 5.6 one may, for example, measure the spatial distribution of a powerful infrared laser field in a plasma by stimulating (with a near-UV range laser) the corresponding SLs of hydrogen-like (e.g., Li III at 2082 Å) or helium-like (e.g., Be III 2080 Å, 2122 Å) ions.

SLs of Non-Coulomb Emitters

The first experiment in which a microwave amplitude was measured by the laser fluorescence technique seems to be the work of the group led by *Kunze* [7.40]. In this (non-plasma) experiment a low-energy lithium beam interacted with microwaves of 9.55 GHz produced by a magnetron. The lithium atoms were excited by simultaneous pulses of two dye lasers. The first pulse at the wavelength $\lambda \approx 6708$ Å populates one of the sublevels of the 2^2P state, i.e., this step has the same result as usually achieved by a collisional excitation in plasmas.

The second pulse was tuned in two different ways. In the first way this pulse was tuned to the wavelength $\lambda \approx 4603$ Å exciting lithium atoms to the level

---

[4] It is worth mentioning here the paper [7.39], the authors of which, after describing their interesting experiment on the diagnostics of microwaves in a plasma by intracavity laser spectroscopy, made a general theoretical conclusion that "unambiguous interpretation of the spectra is possible only when the frequency of the oscillating field exceeds the measured line width". Looking at (7.6.6) one sees that this conclusion is incorrect.

4^2D. Thereafter a fluorescence intensity was registered at the same transition 4^2D $\rightarrow$ 2^2P (4603 Å).

Alternatively, the second pulse was also tuned to a near or far satellite of the 2^2P–4^2F transition. The microwave quanta then act to kick the lithium atoms to the frequency of the 2^2P–4^2F transition. Thereafter a fluorescence intensity was observed at the transition 3^2D $\rightarrow$ 2^2P (6104 Å).

To measure the 4^2 F level population, the fluorescence intensity of the 3^2D $\rightarrow$ 2^2 P transition (6104 Å) was observed. A direct observation of the 4^2F $\rightarrow$ 3^2 D transition is difficult because its wavelength is 1.87 μm. In plasmas, due to collisional mixing of the 4^2 D and 4^2 F states, it is also possible to measure the satellite profiles by looking at the 4^2D $\rightarrow$ 2^2 P transition at 4603 Å. This cannot be done in a collisionless beam experiment, of course.

The measured profiles of the allowed and forbidden line satellites are shown in Fig. 7.26 for two values of the microwave power, $P = 18$ kW and 50 kW. As stated in [7.40], at $P = 18$ kW the satellite intensity ratio S_-/S_+ corresponds to

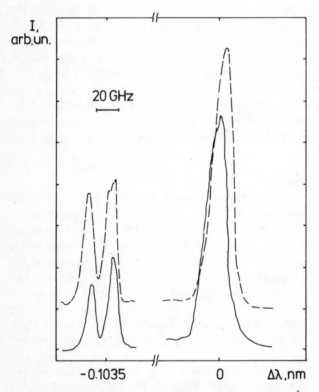

Fig. 7.26. Laser induced fluorescence spectra near the Li I 4603 Å line pumped at 50 kW (*dashed line*) or 18 kW (*solid line*) magnetron power. *Left*: emission intensity at 6104 Å; *right*: intensity at 4603 Å. The allowed line is saturated. From [7.40]

the theoretical results of Baranger and Mozer (Sect. 5.1.1), but at $P = 50$ kW this ratio decreases significantly. The authors note that this cannot be due to saturation [7.40]. This demonstrates the inadequacy of the Dirac perturbation theory for the latter case.

Let us test our adiabatic theory of satellites (Sect. 5.1.2), which is not restricted as is the usual perturbation theory, and analyze the experimental profiles from [7.40]. Calculating the numerical coefficients in (5.1.17) for the separation $\Delta/2\pi c \approx 4.96$ cm^{-1} between the lithium levels 4^2D, 4^2F and substituting into (5.1.17) the experimental ratio S_-/S_+ at $P = 50$ kW, we find a microwave amplitude $E_0 \approx 2.6$ kV/cm.

From the line shifts, the authors of [7.40] estimated the microwave amplitude as $E_0 \approx 2.8$ kV/cm. Thus our method can be successfully used to measure both weak and relatively strong QEFs in plasmas by using SLs with dipole-forbidden components of helium, lithium as well as other noble gases or alkali-like radiators.

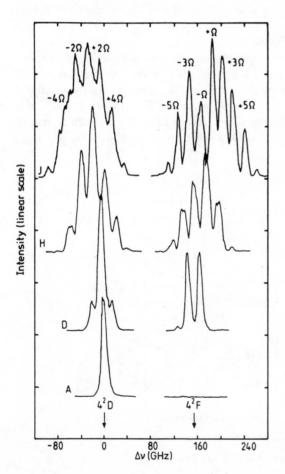

Fig. 7.27. Satellites of 2^2P–4^2D and 2^2P–4^2 F transitions of Li I at a microwave frequency of 9.5 GHz and different microwave amplitudes: A – 0 kV/cm, D – 2.5 kV/cm, H – 8.2 kV/cm, J – 11.5 kV/cm. From [7.41]

Further measurements of microwave fields by laser induced fluorescence of lithium atoms are described in [7.41] and some results are shown in Fig. 7.27. The experiments were performed with the same apparatus as in [7.40]. It is seen that the higher the microwave amplitude the more complicated the satellites structure becomes. Nevertheless, for the entire range of amplitudes used, $0 \leqslant E_0 \leqslant 11.5$ kV/cm, the SL profiles corresponding to the transitions ($2^2P - 4^2D$, 4^2F) of Li can be simulated using our adiabatic theory (Sect. 5.1.2).

It is worth mentioning one more experiment in which laser induced fluorescence was used in a spectroscopic investigation of plasma (rather than maser) satellites in a hollow-cathode discharge operating in helium [7.42]. In this experiment high-voltage pulses of several kV (400 ns duration) were superimposed on a stationary glow discharge of 10 mA total current. Figure 7.28 shows one of the spectra near the SL of He I at 4471 Å at $\Delta t = 190$ ns after the onset of plasma emission. Satellites separated by 120 GHz from the forbidden SL are observed. The most interesting feature is that at the position of the forbidden SL two peaks separated by 22 GHz are seen. It should be noted that all other spectra of the same SL recorded at earlier times exhibit an unsplit forbidden line; the satellite separations from it are smaller than 120 GHz (between 30 and 100 GHz at different $\Delta t \leqslant 150$ ns).

We could suggest the following explanation for these experimental results. At all instants the QEFs (as seen by the behavior of the satellites) as well as quasistatic BEFs (as seen through the forbidden SL) were obviously developed in the plasma. But at the instant $\Delta t = 190$ ns the QEF frequency $\nu = \omega/2\pi = 120$ GHz was equal to half the separation $\Delta\nu = 240$ GHz between forbidden and allowed SLs. In this case a weak satellite of the allowed SL separated

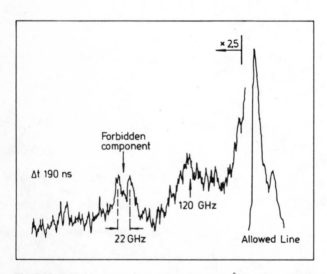

Fig. 7.28. Emission spectrum near the He I 4471 Å line. From [7.42]

from it by 2ν imposes on the forbidden SL and "interacts" with it (in some sense). More precisely, the following situation occurs. Recall that under the joint action of a QEF of some frequency ν and a static EF $\boldsymbol{F}$, in general, both forbidden and allowed SLs may have satellites at multiples of ν (Sect. 6.21). In other words, both the level "f", from which the forbidden SL originates, and the level "a", from which the allowed SL originates, may acquire quasienergy harmonics at frequencies $k\nu$, where $k = \pm1, \pm2, \dots$. In the considered case of $\Delta\nu = 2\nu$ the second quasienergy harmonic of a superimposes on f (and the second quasienergy harmonic of f superimposes on a) and they become degenerate. The nonzero dipole matrix element between the states a and f leads to a mutual repulsion of degenerate QS and their quasienergy harmonics. It is seen as the doublet structure at the position of the forbidden SL (as well as at the position of the allowed SL, which is also seen in Fig. 7.28). Note that an analogous resonance splitting effect under the action of the QEF only (without the quasistatic EF) is described in Sect. 5.2 (see also Fig. 5.8). Thus we must give credit to *Hildebrandt*, who was the first to observe effects of such type [7.42].

It should be noted, however, that most of the theoretical interpretations given by Hildebrandt are believed to be wrong. For example, he asserts that longitudinal Langmuir oscillations, which are characterized only by a scalar electric potential φ at a zero "magnetic" vector-potential A, cannot in principle lead to an appearance of dipole-forbidden SL satellites [7.43]. He based this claim on a formula for the interaction (with an optical electron) of the form $V = -(-e/m_e)\boldsymbol{A}\boldsymbol{d} + A^2 e^2/2m_e$. However, he omitted the term $\Sigma_i e_i \varphi(\boldsymbol{r}_i)$, which may be expanded in a series $\varphi_0 \Sigma_i e_i + (\mathrm{grad}\ \varphi)_0 \Sigma_i e_i \boldsymbol{r}_i + \cdots$, and therefore arrived to the wrong conclusion. This becomes obvious from the consideration of gauge invariance as well.

In a later paper *Hildebrandt*, expressing some doubt about his previous assertions, tried to analyze the problem by solving the Schrödinger equation in a rotating reference frame [7.44]. However, he does not indicate the actual dependence of EFs on the time and also after finding WFs in the rotating reference frame he does not make the return transition to a stationary frame (in which one must finally calculate the spectrum of the atomic radiation). Moreover, a transition into a rotating reference frame works only for emitting particles which can be characterized by a central symmetry potential, and not for helium atoms considered in [7.44].

Perspectives

In all experiments performed up to now on spectroscopic diagnostics of QEFs in plasmas, the measured amplitudes of QEFs were $E_0 \sim 1\text{--}10$ kV/cm.[5] The actual problem is to extend these measurements to lower amplitudes of QEFs. One method would be to use polar molecules as the radiators, which can experience Λ-doubling. As was shown in detail in Sect. 5.5.2 this would allow

---

[5] A review of earlier experiments (before 1976) may be found in [7.45,46].

the limit of applicability to be lowered by two orders of magnitude of QEF amplitude.

Another way to achieve increased sensitivity is to measure the laser absorption at the frequencies of satellites of dipole-forbidden SLs of atoms or ions. This was first proposed by *Kunze* [7.47]. He also pointed out that one might observe this absorption at the frequency of dipole-forbidden SLs as well (because of the presence of quasistatic EFs). This proposal was investigated theoretically in [7.48] and experimentally in a (non-plasma) experiment [7.49]. In this experiment the absorption at the wavelength 4602 Å of the forbidden transition 2^2P–4^2F in lithium was detected by measuring the subsequent cascade fluorescence $4^2F \to 3^2D \to 2^2P$ at the wavelength 6104 Å of the $3^2D \to 2^2P$ transition. It was shown that a static EF as weak as ~ 100 V/cm can be measured by this method. This field strength is one order of magnitude smaller than can be measured by usual methods using the mixing of the same levels 4^2D and 4^2F by EF.

Returning to plasmas containing QEFs, we believe that in order to measure QEF amplitudes as small as $E_0 \gtrsim 10$ V/cm, as is possible by observing the Λ-doublets of polar molecules, one can also use laser absorption at the frequencies of satellites of dipole-forbidden transitions $2\,P$–nF in helium or lithium, where $n = 5$ or 6. Further development of tunable infrared lasers might make it possible to use transitions $3D$–$nG(n = 5, 6)$ of the same atoms that are more sensitive to QEFs.

Appendices

A Tables of Balmer Hydrogen Line Profiles Under the Action of a QEF $E_0 \cos \omega t$

Blochinzew [A.1] tried to find the solution of the Schrödinger equation

$$i\hbar \partial \varphi_n / \partial t = (H_0 + V)\varphi_n, \quad V = ezE_0 \cos \omega t \tag{A.1}$$

in the form

$$\varphi_{n\beta}(\boldsymbol{r}, t) = [\varphi_{n\beta}^{(0)}(\boldsymbol{r}) + u_{n\beta}(\boldsymbol{r}) \exp(i\omega t) + v_{n\beta}(\boldsymbol{r}) \exp(-i\omega t)]\Phi_{n\beta}(t),$$

$$\Phi_{n\beta}(t) = \exp\left\{ -i\mathcal{E}_n t/\hbar - i[3n(n_1 - n_2)_\beta \hbar E_0/2m_e e] \int\limits_0^t d\tau \, \cos \omega \tau \right\}, \tag{A.2}$$

where $H_0 \varphi_{n\beta}^{(0)} = \mathcal{E}_n \varphi_{n\beta}^{(0)}$. The functions $u_{n\beta}$, $v_{n\beta}$ are an order of magnitude smaller, by the parameter $n^2 \hbar^2 E_0 / [m_e e(\mathcal{E}_n - \mathcal{E}_{n'} \pm \hbar\omega)]$, than $\varphi_{n\beta}^{(0)}$. Therefore, the spontaneous emission spectrum is determined by a "reduced" WF $\varphi_{n\beta}^{(0)} \Phi_{n\beta}(t)$ or, more precisely, by the Fourier expansion of

$$\Phi_{n\beta}^*(t)\Phi_{n'\beta}(t) = \exp(iX\varepsilon \sin \omega t),$$

$$\varepsilon \equiv 3\hbar E_0 / 2m_e e\omega, \quad X \equiv n(n_1 - n_2) - n'(n_1' - n_2'). \tag{A.3}$$

This Fourier expansion has the form

$$\exp(iX\varepsilon \sin \omega t) = \sum_{p=-\infty}^{+\infty} J_p(X\varepsilon) \exp(ip\omega t), \tag{A.4}$$

where $J_P(z)$ are Bessel functions. Thus one obtains (3.1.1 and 9) for the hydrogen line profile in the hypothetical (one-component) case and in the real (multicomponent) case correspondingly.

We have calculated the line profiles $S_B(\Delta\omega/\omega, \varepsilon)$ using (3.1.9) for observations in the direction orthogonal to E_0. In the following tables, the satellite intensities $I(p, \varepsilon)$ are given for each value of $\varepsilon \equiv 3\hbar E_0/2m_e e\omega$ beginning with $p = 0$: $I(0, \varepsilon)$, $I(1, \varepsilon)$, $I(2, \varepsilon)$, ..., $I(p_{\text{last}}, \varepsilon)$. The calculations spanned satellite intensities (from the maximum to the wing) over four orders of magnitude: $I(p_{\text{last}}, \varepsilon)/ \max\limits_p I(p, \varepsilon) < 10^{-4}$. At each ε the line profile is normalized to unity: $I(0, \varepsilon) + 2\sum_{p=1}^{\infty} I(p, \varepsilon) = 1$. The halfhalfwidth $p_{1/2} = \Delta\omega_{1/2}^{\text{half}}/\omega$ of each profile is also given.

To allow for additional broadening mechanisms one can use (3.1.9) for $S(\Delta\omega/\omega)$, replacing $\delta(\Delta\omega/\omega - p)$ by the function corresponding to the line shape of the broadening mechanism. For example, for Doppler broadening we obtain

$$S_{B+D}\left(\frac{\Delta\omega}{\omega}\right) = \sum_{p=-\infty}^{+\infty} I(p, \varepsilon) \int_{-\infty}^{+\infty} \delta\left(\frac{\Delta\omega}{\omega} - p - \frac{v\omega_0}{c\omega}\right) f\left(\frac{v}{v_0}\right) d\left(\frac{v}{v_0}\right)$$

$$= \sum_{p=-\infty}^{+\infty} I(p, \varepsilon) \frac{c\omega}{v_0\omega_0} F\left(\frac{(\Delta\omega - p\omega)c}{v_0\omega_0}\right), \tag{A.5}$$

where $F(x) = \pi^{-1/2} \exp(-x^2)$, $v_0 = (2T/M)^{1/2}$. The values of $I(p, \varepsilon)$ tabulated below allow direct calculation using (A.5).

Table A.1. Balmer hydrogen line profiles in a QEF $E_0 \cos \omega t$

H_α LINE

ε	$p_{1/2}$	$I(p, \varepsilon)$				
0.10	0.5072	9.722E-01	1.377E-02	1.080E-04	4.485E-07	1.255E-09
		2.950E-12	6.831E-15	1.569E-17	3.295E-20	
0.20	0.5296	8.964E-01	5.017E-02	1.604E-03	2.686E-05	2.998E-07
		2.782E-09	2.542E-11	2.325E-13	1.959E-15	
0.30	0.5696	7.919E-01	9.660E-02	7.175E-03	2.742E-04	6.864E-06
		1.404E-07	2.820E-09	5.747E-11	1.093E-12	
0.40	0.6267	6.829E-01	1.381E-01	1.909E-02	1.326E-03	5.889E-05
		2.087E-06	7.190E-08	2.560E-09	8.683E-11	
0.50	0.6908	5.903E-01	1.630E-01	3.737E-02	4.184E-03	2.912E-04
		1.566E-05	8.041E-07	4.344E-08	2.301E-09	
0.60	0.7333	5.253E-01	1.672E-01	5.914E-02	9.951E-03	1.008E-03
		7.589E-05	5.294E-06	3.935E-07	2.977E-08	
0.70	0.7297	4.888E-01	1.538E-01	7.954E-02	1.924E-02	2.705E-03
		2.716E-04	2.419E-05	2.293E-06	2.309E-07	
0.80	0.6926	4.728E-01	1.315E-01	9.360E-02	3.161E-02	5.992E-03
		7.783E-04	8.508E-05	9.675E-06	1.218E-06	
0.90	0.6535	4.664E-01	1.096E-01	9.833E-02	4.537E-02	1.137E-02
		1.878E-03	2.468E-04	3.223E-05	4.774E-06	
1.00	0.6296	4.599E-01	9.467E-02	9.382E-02	5.795E-02	1.894E-02
		3.936E-03	6.180E-04	9.043E-05	1.492E-05	
1.10	0.6236	4.483E-01	8.888E-02	8.309E-02	6.673E-02	2.817E-02
		7.318E-03	1.372E-03	2.240E-04	3.953E-05	
1.30	0.6483	4.126E-01	9.438E-02	6.022E-02	6.793E-02	4.629E-02
		1.858E-02	4.996E-03	1.043E-03	2.044E-04	4.570E-05
1.50	0.6781	3.794E-01	9.964E-02	5.157E-02	5.474E-02	5.398E-02
		3.292E-02	1.282E-02	3.558E-03	8.195E-04	1.885E-04
		5.056E-05				

Table A.1. *Continued*

H_α LINE

ε	$p_{1/2}$	$I(p,\varepsilon)$				
1.70	0.6956	3.544E-01	9.966E-02	5.090E-02	4.426E-02	4.892E-02
		4.235E-02	2.402E-02	9.150E-03	2.615E-03	6.623E-04
		1.762E-04	5.428E-05			
1.90	0.7144	3.322E-01	9.969E-02	5.060E-02	3.922E-02	4.168E-02
		4.199E-02	3.334E-02	1.792E-02	6.703E-03	1.968E-03
		5.483E-04	1.676E-04	5.742E-05		
2.10	0.6993	3.210E-01	9.149E-02	5.695E-02	3.241E-02	3.826E-02
		3.754E-02	3.547E-02	2.651E-02	1.362E-02	5.008E-03
		1.508E-03	4.629E-04	1.623E-04	6.073E-05	
2.60	0.6298	3.160E-01	6.513E-02	5.961E-02	4.328E-02	1.859E-02
		3.123E-02	3.335E-02	2.726E-02	2.564E-02	2.006E-02
		1.107E-02	4.440E-03	1.444E-03	4.927E-04	2.216E-04
		1.087E-04	4.748E-05	2.044E-05	1.138E-05	7.667E-06
		4.785E-06	2.494E-06	1.079E-06	3.934E-07	1.232E-07
		3.365E-08				
3.10	0.5682	3.209E-01	3.852E-02	6.885E-02	2.834E-02	3.997E-02
		1.480E-02	1.868E-02	3.315E-02	2.542E-02	1.974E-02
		1.992E-02	1.624E-02	9.438E-03	4.069E-03	1.402E-03
		4.750E-04	2.303E-04	1.490E-04	8.847E-05	4.187E-05
		1.719E-05	9.017E-06	7.204E-06	5.879E-06	3.982E-06
		2.202E-06	1.016E-06	4.002E-07	1.370E-07	4.136E-08
3.60	0.5202	3.415E-01	1.325E-02	5.550E-02	5.436E-02	1.174E-02
		3.307E-02	2.086E-02	7.109E-03	2.645E-02	2.946E-02
		1.775E-02	1.490E-02	1.649E-02	1.377E-02	8.272E-03
		3.792E-03	1.395E-03	4.639E-04	2.075E-04	1.516E-04
		1.181E-04	7.692E-05	3.942E-05	1.600E-05	6.796E-06
		5.401E-06	5.661E-06	4.981E-06	3.485E-06	1.997E-06
		9.656E-07	4.033E-07	1.480E-07	4.833E-08	
4.10	0.5263	3.331E-01	1.667E-02	3.571E-02	5.055E-02	3.710E-02
		8.633E-03	2.274E-02	2.653E-02	5.456E-03	1.378E-02
		2.969E-02	2.273E-02	1.199E-02	1.196E-02	1.431E-02
		1.209E-02	7.382E-03	3.540E-03	1.409E-05	4.818E-04
		1.819E-04	1.287E-04	1.221E-04	1.005E-04	6.792E-05
		3.760E-05	1.642E-05	5.893E-06	3.367E-06	4.221E-06
		4.937E-06	4.413E-06	3.129E-06	1.838E-06	9.223E-07
		4.039E-07	1.568E-07	5.460E-08		
4.60	0.5423	3.205E-01	2.500E-02	1.869E-02	4.717E-02	3.859E-02
		2.303E-02	1.043E-02	1.594E-02	2.525E-02	1.085E-02
		5.080E-03	2.160E-02	2.738E-02	1.595E-02	8.242E-03
		1.023E-02	1.279E-02	1.088E-02	6.710E-03	3.293E-03
		1.398E-03	5.300E-04	1.844E-04	9.476E-05	4.495E-06
		3.997E-06	2.854E-06	1.710E-06	8.843E-07	4.028E-07
		6.457E-06	2.227E-06	2.344E-06	3.796E-06	4.495E-06
		3.997E-06	2.854E-06	1.710E-06	8.843E-07	4.028E-07
		1.639E-07	6.021E-08			
5.10	0.5645	3.045E-01	3.478E-02	9.586E-03	3.585E-02	4.222E-02
		2.730E-02	1.550E-02	1.027E-02	1.464E-02	2.037E-02
		1.466E-02	4.573E-03	1.154E-02	2.488E-02	2.228E-02
		1.049E-02	6.030E-03	9.212E-03	1.167E-02	9.949E-03

Table A.1. *Continued*

H_α LINE

ε	$p_{1/2}$	$I(p,\varepsilon)$				
		6.201E-03	3.080E-03	1.348E-03	5.646E-04	2.216E-04
		8.586E-05	6.564E-05	8.816E-05	9.803E-05	8.164E-05
		5.462E-05	3.209E-05	1.718E-05	7.677E-06	2.417E-06
		1.029E-06	2.108E-06	3.599E-06	4.166E-06	3.667E-06
		2.631E-06	1.602E-06	8.505E-07	4.006E-07	1.696E-07
		6.524E-08				

H_β LINE

ε	$p_{1/2}$	$I(p,\varepsilon)$				
0.05	0.5161	9.395E-01	2.992E-02	3.410E-04	2.042E-06	7.527E-09
		1.894E-11	3.519E-14	5.144E-17	6.250E-20	
0.10	0.5772	7.809E-01	1.045E-01	4.900E-03	1.196E-04	1.787E-06
		1.814E-08	1.356E-10	7.952E-13	3.870E-15	
0.20	1.2271	3.997E-01	2.436E-01	5.080E-02	5.361E-03	3.379E-04
		1.422E-05	4.348E-07	1.034E-08	2.025E-10	
0.30	2.0674	2.047E-01	2.350E-01	1.236E-01	3.334E-02	5.192E-03
		5.243E-04	3.759E-05	2.059E-06	9.177E-08	
0.40	2.8780	1.497E-01	1.730E-01	1.428E-01	7.867E-02	2.497E-02
		4.946E-03	6.730E-04	6.817E-05	5.503E-06	
0.50	3.5434	1.005E-01	1.508E-01	1.189E-01	9.777E-02	5.660E-02
		2.010E-02	4.710E-03	7.927E-04	1.032E-04	1.111E-05
0.60	4.3548	7.492E-02	1.252E-01	1.127E-01	8.544E-02	7.292E-02
		4.382E-02	1.693E-02	4.500E-03	8.904E-04	1.409E-04
		1.905E-05				
0.70	5.0969	7.669E-02	9.035E-02	1.108E-01	8.137E-02	6.573E-02
		5.755E-02	3.561E-02	1.470E-02	4.314E-03	9.707E-04
		1.799E-04	2.935E-05			
0.80	6.0200	7.125E-02	7.412E-02	9.374E-02	8.604E-02	6.170E-02
		5.311E-02	4.722E-02	2.994E-02	1.304E-02	4.148E-03
		1.037E-03	2.191E-04	4.188E-05	7.549E-06	
0.90	6.6389	6.984E-02	5.961E-02	8.296E-02	7.908E-02	6.697E-02
		4.923E-02	4.435E-02	3.986E-02	2.583E-02	1.176E-02
		3.998E-03	1.091E-03	2.575E-04	5.638E-05	1.189E-05
1.00	7.1643	7.061E-02	4.778E-02	7.263E-02	7.464E-02	6.360E-02
		5.417E-02	4.072E-02	3.789E-02	3.441E-02	2.272E-02
		1.074E-02	3.862E-03	1.134E-03	2.942E-04	7.245E-05
		1.751E-05				
1.30	8.4613	3.726E-02	5.389E-02	3.930E-02	5.493E-02	6.018E-02
		5.339E-02	3.915E-02	3.992E-02	3.372E-02	2.584E-02
		2.597E-02	2.432E-02	1.681E-02	8.651E-03	3.519E-03
		1.215E-03	3.883E-04	1.247E-04	4.150E-05	1.377E-05
		4.302E-06				
1.60	7.7135	4.594E-02	2.135E-02	6.612E-02	7.013E-02	6.159E-02
		4.834E-02	3.592E-02	4.579E-02	2.811E-02	2.666E-02
		3.107E-02	2.295E-02	1.787E-02	1.965E-02	1.893E-02
		1.348E-02	7.341E-03	3.251E-03	1.251E-03	4.553E-04
		1.727E-04	7.097E-05	3.042E-05	1.253E-05	4.706E-06

Table A.1. *Continued*

H_β LINE

ε	$p_{1/2}$	$I(p, \varepsilon)$				
1.90	9.5807	3.248E-02	2.926E-02	2.778E-02	4.941E-02	1.673E-02
		2.998E-02	6.448E-02	2.847E-02	2.182E-02	4.147E-02
		2.557E-02	1.530E-02	2.534E-02	2.490E-02	1.544E-02
		1.314E-02	1.601E-02	1.570E-02	1.136E-02	6.422E-03
		3.029E-03	1.267E-03	5.042E-04	2.090E-04	9.722E-05
		4.972E-05	2.547E-05	1.212E-05	5.171E-06	
2.10	11.2997	2.242E-02	3.328E-02	1.995E-02	4.058E-02	3.115E-02
		1.690E-02	3.531E-02	5.435E-02	3.038E-02	1.308E-02
		3.356E-02	3.318E-02	1-313E-02	1.511E-02	2.531E-02
		2.062E-02	1.171E-02	1.114E-02	1.440E-02	1.421E-02
		1.034E-02	5.941E-03	2.893E-03	1.270E-03	5.330E-04
		2.301E-04	1.110E-04	6.087E-05	3.492E-05	1.904E-05
		9.387E-06	4.115E-06	1.604E-06	5.590E-07	1.754E-07
		4.987E-08	1.294E-08	3.077E-09	6.749E-10	
2.40	17.0321	2.427E-02	2.377E-02	2.490E-02	2.371E-02	3.372E-02
		2.692E-02	1.425E-02	3.054E-02	4.543E-02	3.458E-02
		1.770E-02	1.405E-02	2.775E-02	3.033E-02	1.374E-02
		7.858E-03	1.895E-02	2.299E-02	1.444E-02	7.775E-03
		9.252E-03	1.271E-02	1.255E-02	9.173E-03	5.357E-03
		2.696E-03	1.257E-03	5.721E-04	2.633E-04	1.294E-04
		7.294E-05	4.343E-05	3.000E-05	1.800E-05	9.657E-06
		4.595E-06	1.947E-06	7.400E-07	2.542E-07	7.946E-08
		2.275E-08	6.000E-09	1.464E-09	3.320E-10	
2.70	19.8542	2.161E-02	2.171E-02	1.991E-02	2.495E-02	2.105E-02
		3.160E-02	2.051E-02	1.455E-02	2.643E-02	3.773E-02
		3.664E-02	1.962E-02	1.154E-02	1.714E-02	2.437E-02
		2.549E-02	1.506E-02	5.805E-03	1.155E-02	2.084E-02
		1.853E-02	9.417E-03	5.446E-03	8.213E-03	1.154E-02
		1.133E-02	8.311E-03	4.906E-03	2.513E-03	1.218E-03
		5.978E-04	3.002E-04	1.528E-04	8.328E-05	5.287E-05
		3.774E-05	2.675E-05	1.731E-05	9.950E-06	5.067E-06
		2.302E-06	9.407E-07	3.486E-07	1.180E-07	3.373E-08
		1.056E-08	2.820E-09	7.020E-10		
3.00	22.2591	4.184E-03	3.383E-02	6.677E-03	2.849E-02	1.702E-02
		2.385E-02	2.556E-02	1.791E-02	1.472E-02	2.101E-02
		3.601E-02	3.193E-02	2.349E-02	1.322E-02	9.320E-03
		1.846E-02	2.346E-02	2.035E-02	1.406E-02	6.964E-03
		7.146E-03	1.541E-02	1.953E-02	1.365E-02	5.850E-03
		4.210E-03	7.646E-03	1.066E-02	1.036E-02	7.638E-03
		4.560E-03	2.356E-03	1.158E-03	5.992E-04	3.318E-04
		1.841E-04	9.987E-05	5.816E-05	4.058E-05	3.188E-05
		2.440E-05	1.682E-05	1.025E-05	5.527E-06	2.665E-06
		1.159E-06	4.586E-07	1.663E-07	5.557E-08	1.722E-08
		4.967E-09	1.339E-09	3.387E-10	8.059E-11	
3.50	26.4034	1.747E-02	1.578E-02	1.609E-02	2.020E-02	9.681E-03
		2.693E-02	1.840E-02	1.192E-02	3.072E-02	7.041E-03
		1.088E-02	3.066E-02	2.260E-02	2.975E-02	2.519E-02
		7.942E-03	1.045E-02	1.279E-02	1.293E-02	2.062E-02
		2.024E-02	1.131E-02	7.145E-03	7.001E-03	7.857E-03

Table A.1. *Continued*

H_β LINE

ε	$p_{1/2}$	$I(p, \varepsilon)$				
		1.205E-02	1.623E-02	1.408E-02	7.178E-03	2.577E-03
		3.496E-03	7.180E-03	9.565E-03	9.111E-03	6.768E-03
		4.131E-03	2.171E-03	1.062E-03	5.541E-04	3.420E-04
		2.302E-04	1.464E-04	8.364E-05	4.690E-05	3.169E-05
		2.713E-05	2.435E-05	1.994E-05	1.434E-05	9.073E-06
		5.107E-06	2.586E-06	1.190E-06	5.014E-07	1.949E-07
		7.027E-08	2.361E-08	7.427E-09	2.194E-09	6.108E-10
4.00	24.0811	1.493E-02	1.387E-02	1.504E-02	1.398E-02	1.646E-02
		1.195E-02	1.868E-02	2.178E-02	1.082E-02	1.575E-02
		2.142E-02	8.579E-03	9.324E-03	2.511E-02	2.984E-02
		1.966E-02	1.773E-02	1.915E-02	6.150E-03	3.085E-03
		1.540E-02	1.641E-02	1.255E-02	1.698E-02	1.570E-02
		6.075E-03	2.995E-03	7.174E-03	9.577E-03	1.033E-02
		1.252E-02	1.299E-02	8.828E-03	3.330E-03	1.345E-03
		3.528E-03	6.934E-03	8.726E-03	8.155E-03	6.101E-03
		3.807E-03	2.048E-03	1.001E-03	5.017E-04	3.066E-04
		2.320E-04	1.820E-04	1.281E-04	7.699E-05	4.135E-05
		2.456E-05	2.063E-05	2.107E-05	2.039E-05	1.724E-05
		1.268E-05	8.231E-06	4.781E-06	2.515E-06	1.210E-06
		5.364E-07	2.205E-07	8.452E-08	3.034E-08	1.023E-08
		3.256E-09	9.797E-10	2.795E-10		
4.50	25.0763	7.193E-03	1.848E-02	7.194E-03	1.816E-02	9.306E-03
		1.575E-02	1.160E-02	1.594E-02	1.842E-02	1.451E-02
		8.796E-03	1.720E-02	1.593E-02	7.660E-03	1.072E-02
		2.150E-02	2.869E-02	2.000E-02	1.288E-02	1.446E-02
		1.190E-02	5.885E-03	3.960E-03	1.085E-02	1.824E-02
		1.472E-02	9.727E-03	1.247E-02	1.201E-02	4.012E-03
		8.813E-04	6.424E-03	1.098E-02	1.043E-02	9.633E-03
		1.044E-02	9.380E-03	5.218E-03	1.372E-03	1.055E-03
		3.762E-03	6.761E-03	8.046E-03	7.373E-03	5.560E-03
		3.556E-03	1.962E-03	9.646E-04	4.656E-04	2.675E-04
		2.043E-04	1.801E-04	1.537E-04	1.160E-04	7.424E-05
		4.018E-05	2.066E-05	1.458E-05	1.579E-05	1.803E-05
		1.799E-05	1.539E-05	1.147E-05	7.581E-06	4.515E-06
		2.450E-06	1.223E-06	5.652E-07	2.434E-07	9.815E-08
		3.720E-08	1.330E-08	4.501E-09	1.445E-09	4.412E-10
		1.284E-10				
5.00	28.4048	1.295E-02	1.022E-02	1.236E-02	1.213E-02	8.839E-03
		1.840E-02	3.103E-03	1.974E-02	1.141E-02	1.260E-02
		2.152E-02	2.037E-03	1.517E-02	1.535E-02	9.659E-03
		1.157E-02	7.542E-03	2.165E-02	2.655E-02	1.773E-02
		1.420E-02	9.482E-03	1.007E-02	1.085E-02	4.009E-03
		3.967E-03	1.144E-02	1.544E-02	1.484E-02	1.097E-02
		7.677E-03	8.960E-03	9.041E-03	3.576E-03	4.157E-04
		5.192E-03	1.114E-02	1.136E-02	8.512E-03	7.666E-03
		8.152E-03	6.600E-03	3.002E-03	5.027E-04	1.157E-03
		3.979E-03	6.592E-03	7.502E-03	6.731E-03	5.086E-03
		3.334E-03	1.909E-03	9.566E-04	4.415E-04	2.297E-04

Table A.1. *Continued*

H_β LINE

ε	$p_{1/2}$	$I(p, \varepsilon)$				
		1.714E-04	1.616E-04	1.530E-04	1.339E-04	1.052E-04
		7.193E-05	4.138E-05	2.020E-05	1.075E-05	1.029E-05
		1.362E-05	1.628E-05	1.634E-05	1.401E-05	1.052E-05
		7.055E-06	4.289E-06	2.389E-06	1.230E-06	5.890E-07
		2.638E-07	1.111E-07	4.409E-08	1.656E-08	5.906E-09
		2.003E-09	6.478E-10	2.001E-10		
6.00	45.9772	8.342E-03	1.095E-02	8.059E-03	1.139E-02	8.197E-03
		1.084E-02	9.416E-03	1.085E-02	9.483E-03	8.955E-03
		1.560E-02	1.218E-02	8.021E-03	1.369E-02	9.361E-03
		5.847E-03	1.124E-02	1.577E-02	7.404E-03	3.914E-03
		1.470E-02	1.938E-02	1.811E-02	1.796E-02	1.474E-02
		6.934E-03	4.084E-03	1.017E-02	9.687E-03	2.638E-03
		4.423E-03	8.370E-03	7.635E-03	1.164E-02	1.600E-02
		1.068E-02	4.616E-03	5.870E-03	7.429E-03	5.408E-03
		3.918E-03	3.433E-03	2.531E-03	3.910E-03	8.531E-03
		1.158E-02	9.644E-03	5.699E-03	4.126E-03	4.947E-03
		5.173E-03	3.344E-03	9.325E-04	1.653E-04	1.672E-03
		4.240E-03	6.185E-03	6.645E-03	5.804E-03	4.354E-03
		2.910E-03	1.771E-03	9.830E-04	4.844E-04	2.077E-04
		9.541E-05	8.566E-05	1.138E-04	1.334E-04	1.293E-04
		1.092E-04	8.530E-05	6.300E-05	4.270E-05	2.478E-05
		1.145E-05	4.781E-06	4.502E-06	8.007E-06	1.202E-05
		1.425E-05	1.406E-05	1.200E-05	9.100E-06	6.240E-06
		3.920E-06	2.277E-06	1.232E-06	6.249E-07	2.983E-07
		1.346E-07	5.760E-08	2.344E-08	9.095E-09	3.372E-09
		1.197E-09	4.075E-10	1.333E-10	4.193E-11	1.271E-11
		3.714E-12	1.048E-12			
8.00	45.8322	1.268E-02	1.657E-03	1.289E-02	1.447E-03	1.290E-02
		2.354E-03	1.070E-02	6.491E-03	5.654E-03	1.124E-02
		3.324E-03	1.276E-02	1.969E-03	1.189E-02	1.068E-02
		7.420E-03	9.726E-03	7.721E-03	7.329E-03	4.254E-03
		1.133E-02	7.359E-03	2.864E-03	9.569E-03	1.189E-02
		8.186E-03	2.920E-03	7.374E-03	1.312E-02	1.397E-02
		1.736E-02	1.482E-02	7.514E-03	6.200E-03	7.978E-03
		6.114E-03	1.986E-03	3.462E-03	7.829E-03	6.185E-03
		3.910E-03	4.759E-03	3.194E-03	4.031E-03	1.069E-02
		1.254E-02	7.900E-03	6.618E-03	8.022E-03	5.931E-03
		2.953E-03	2.576E-03	2.928E-03	3.572E-03	5.142E-03
		4.857E-03	1.693E-03	1.861E-04	3.180E-03	7.056E-03
		7.854E-03	6.708E-03	6.431E-03	6.748E-03	5.587E-03
		2.920E-03	9.846E-04	1.288E-03	2.846E-03	3.554E-03
		2.546E-03	8.335E-04	1.096E-05	7.626E-04	2.548E-03
		4.322E-03	5.323E-03	5.352E-03	4.625E-03	3.519E-03
		2.389E-03	1.475E-03	8.576E-04	4.954E-04	2.909E-04
		1.630E-04	7.516E-05	2.358E-05	1.094E-05	2.973E-05
		6.249E-05	9.104E-05	1.045E-04	1.012E-04	8.597E-05
		6.559E-05	4.566E-05	2.933E-05	1.753E-05	9.816E-06
		5.174E-06	2.579E-06	1.220E-06	5.492E-07	2.360E-07
		9.698E-08	3.820E-08	1.445E-08	5.254E-09	1.840E-09

B Reduced Halfhalfwidths $\alpha_{1/2} \equiv \Delta\lambda_{1/2}/F_0$ of Hydrogen and Ionized Helium Spectral Lines in a Linearly Polarized Multimode QEF of rms Amplitude F_0 for Transverse (t) and Longitudinal (l) Observations

$Z = 1, n' = 2$						
n	3	4	5	6	7	8
$(\alpha_{1/2})_t \times 10^2$	1.428	2.584	2.644	2.912	4.400	3.565
$(\alpha_{1/2})_l \times 10^2$	0.984	2.069	2.222	2.200	4.017	2.585

$Z = 2, n' = 3$				$Z = 2, n' = 4$		
n	4	5	6	n	5	6 · 7
$(\alpha_{1/2})_t \times 10^2$	0.326	0.717	0.589	$(\alpha_{1/2})_t \times 10^2$	1.713	3.725 1.668
$(\alpha_{1/2})_l \times 10^2$	0.262	0.577	0.399	$(\alpha_{1/2})_l \times 10^2$	1.342	2.952 1.559

C Fourier Coefficients of an EF with Nonorthogonal Components F and $E_0 \cos \omega t$

The coefficients b_r in (4.2.10) are equal to:

$$b_1 = FE_{0x}E_0^{-2}(G^{(+)} + G^{(-)} - 2)\omega,$$

$$b_2 = \frac{FE_{0x}}{E_0^2}\left[\left(1 - \frac{2FE_{0z}}{E_0^2}\right)G^{(+)} - \left(1 + \frac{2FE_{0z}}{E_0^2}\right)G^{(-)} + \frac{4FE_{0z}}{E_0^2}\right]\omega,$$

$$b_3 = \frac{FE_{0x}}{E_0^2}\left[6 - \frac{32F^2E_{0z}^2}{E_0^4} + \frac{8F^2}{E_0^2}\right.$$

$$- \left(1 + \frac{4F^2}{E_0^2} + \frac{8FE_{0z}}{E_0^2} - \frac{16F^2E_{0z}^2}{E_0^4}\right)G^{(+)}$$

$$\left. - \left(1 + \frac{4F^2}{E_0^2} - \frac{8FE_{0z}}{E_0^2} - \frac{16F^2E_{0z}^2}{E_0^4}\right)G^{(-)}\right]\omega,$$

$$G^{(\pm)} = \left(\frac{2(F^2 + E_0^2 \pm 2FE_{0z})}{F^2 - E_0^2 + [(F^2 + E_0^2)^2 - 4F^2E_{0z}^2]^{1/2}}\right)^{1/2}.$$

The coefficients ε_j in (4.2.10) may be represented in the form:

$$\varepsilon_0 = (F^2 + E_0^2)^{1/2}\left(\alpha_0^{(0)} - \frac{1 \cdot 1}{2 \cdot 4}f^2\alpha_0^{(2)} - \cdots\right)$$

$$\varepsilon_1 = (F^2 + E_0^2)^{1/2}\left(\frac{1}{2}f\alpha_1^{(1)} + \frac{1 \cdot 1 \cdot 3}{2 \cdot 4 \cdot 6}f^3\alpha_1^{(3)} + \cdots\right)$$

$$\varepsilon_2 = (F^2 + E_0^2)^{1/2} \left(\alpha_2^{(0)} - \frac{1 \cdot 1}{2 \cdot 4} f^2 \alpha_2^{(2)} - \cdots \right)$$

$$\varepsilon_3 \approx (F^2 + E_0^2)^{1/2} f \alpha_3^{(1)} / 2,$$

$$\varepsilon_4 \approx (F^2 + E_0^2)^{1/2} \alpha_4^{(0)}$$

$$\alpha_0^{(0)} = 4\mathbb{E}(k)/\pi, \quad \alpha_0^{(2)} = 4\mathbb{D}(k)/\pi$$

$$\alpha_1^{(1)} = (4/\pi)[\mathbb{K}(k) - \mathbb{D}(k)],$$

$$\alpha_1^{(3)} = [4/(3\pi k^2)][-\mathbb{E}(k) + (2 + k^2)\mathbb{D}(k)]$$

$$\alpha_2^{(0)} = [4/(3\pi)][\mathbb{E}(k) - 2(1 - k^2)\mathbb{D}(k)],$$

$$\alpha_2^{(2)} = [4/(\pi k^2)][2\mathbb{E}(k) - (4 - 3k^2)\mathbb{D}(k)]$$

$$\alpha_3^{(1)} = [4/(3\pi k^4)][(7k^2 - 8)\mathbb{E}(k) + (3k^4 - 11k^2 + 8)\mathbb{K}(k)]$$

$$\alpha_4^{(0)} = [4/(15\pi k^4)][(8k^4 - 24k^2 + 16)\mathbb{K}(k) + (-k^4 + 16k^2 - 16)\mathbb{E}(k)]$$

$$k^2 \equiv E_0^2/(F^2 + E_0^2), \quad f \equiv 2FE_{0z}/(F^2 + E_0^2)$$

$\mathbb{E}(k)$, $\mathbb{D}(k)$, $\mathbb{K}(k)$ are complete elliptic integrals.

D Generalized QSs of a Hydrogen Atom in a Bichromatic EF

Consider the Schrödinger equation for a hydrogen atom in an EF (3.4.1)

$$i \, \partial \psi / \partial t = [H_0 + r E_{01} \cos(\omega_1 t + \theta) + z E_{02} \cos \omega_2 t] \psi \qquad \text{(D.1)}$$

at the conditions of resonance (3.4.2). We try its solution in the form

$$\psi(t) = \sum_{\substack{\alpha \in a \\ \beta \in b}} [\exp(it\delta/2) A_\alpha(t) \psi_\alpha(t) + \exp(-it\delta/2) B_\beta(t) \psi_\beta(t)],$$

$$\psi_\mu(\mathbf{r}, t) = \varphi_\mu(\mathbf{r}) \exp(-iE_\mu^{(0)} - i z_{\mu\mu} \omega_2^{-1} E_{02} \sin \omega_2 t), \quad \mu = \alpha, \beta; \qquad \text{(D.2)}$$

where $\varphi_\mu(\mathbf{r})$ are eigen-WFs of H_0 in the parabolic coordinates and $E_\mu^{(0)}$ are the unperturbed energies. Substituting $\psi(t)$ in (D.1) we obtain the system of equations

$$i\dot{A}_{\alpha'} = \delta A_{\alpha'}/2 + \cos(\omega_1 t + \theta) \sum_{\substack{\alpha \in a \\ \beta \in b}} [A_\alpha \langle \psi_{\alpha'}(t) | \mathbf{r} E_{01} | \psi_\alpha(t) \rangle$$

$$+ B_\beta \langle \psi_{\alpha'}(t) | \mathbf{r} E_{01} | \psi_\beta(t) \rangle \exp(-it\delta)],$$

$$i\dot{B}_{\beta'} = -\delta B_{\beta'}/2 + \cos(\omega_1 t + \theta) \sum_{\substack{\alpha \in a \\ \beta \in b}} [\exp(it\delta) A_\alpha \langle \psi_{\beta'}(t) | \mathbf{r} E_{01} | \psi_\alpha(t) \rangle$$

$$+ B_\beta \langle \psi_{\beta'}(t) | \mathbf{r} E_{01} | \psi_\beta(t) \rangle], \quad \alpha' \in a, \quad \beta' \in b. \qquad \text{(D.3)}$$

In the resonance approximation, keeping in (D.3) only the terms with constant coefficients (this corresponds to finding the quasienergies up to the terms linear in E_{01} at $\delta = 0$), we obtain

$$i\dot{A}_{\alpha'} = \frac{\delta}{2}A_{\alpha'} + \frac{1}{2}\exp(-i\theta)\sum_{\beta\in b}B_\beta J_r\left(\frac{\Delta z_{\alpha'\beta}E_{02}}{\omega_2}\right)(rE_{01})_{\alpha'\beta}, \quad \alpha' \in a,$$

$$i\dot{B}_{\beta'} = -\frac{\delta}{2}B_{\beta'} + \frac{1}{2}\exp(i\theta)\sum_{\alpha\in a}A_\alpha J_r\left(\frac{\Delta z_{\alpha\beta'}E_{02}}{\omega_2}\right)(rE_{01})_{\alpha\beta'}, \quad \beta' \in b,$$

$$\Delta z_{\alpha\beta} \equiv z_{\alpha\alpha} - z_{\beta\beta}, (rE_{01})_{\alpha\beta} \equiv \langle\varphi_\alpha|rE_{01}|\varphi_\beta\rangle. \tag{D.4}$$

The neglect of terms with oscillatory coefficients in (D.3) is justified when

$$\left|\frac{(rE_{01})_{\alpha\beta}J_p(\Delta z_{\alpha\beta}E_{02}/\omega_2)}{2(p-r)\omega_2}\right| \ll 1, \quad p \neq r; \quad \alpha \in a, \quad \beta \in b. \tag{D.5}$$

The solution of (D.4) corresponding to the given quasienergies we represent in the form

$$X_j = X_j(0)\exp(\kappa_j t), \tag{D.6}$$

where

$$X_j = (A_1^{(j)}, \ldots, A_{n_a^2}^{(j)}, B_1^{(j)}, \ldots, B_{n_b^2}^{(j)})^T,$$

$$X_j(0) = (A_1^{(j)}(0), \ldots, A_{n_a^2}^{(j)}(0), B_1^{(j)}(0), \ldots, B_{n_b^2}^{(j)}(0))^T, \tag{D.7}$$

and T is the transposition operator.

For Lyman lines ($n_b = 1$), from (D.7) it is easy to obtain

$$\kappa_{1,2} = \pm i\Omega_0(\delta, r), \qquad \kappa_{3,4,\ldots,n_a^2+1} = -i\delta/2, \tag{D.8}$$

with the frequency of quantum oscillations (generalized Rabi frequency) $\Omega_0(\delta, r)$ being equal to

$$\Omega_0(\delta, r) = 2^{-1}\left[\delta^2 + \sum_{\alpha=1}^{n_a^2}|(rE_{01})_{\beta\alpha}|^2 J_r(z_{\alpha\alpha}E_{02}/\omega_2)\right]^{1/2}. \tag{D.9}$$

For the H_α line the values of κ_j (when $E_{01} \parallel E_{02}$) are

$$\kappa_{1,2,3,4,5} = -i\delta/2, \qquad \kappa_{6,7} = \pm\Omega_1(\delta, r), \qquad \kappa_{8,9} = \pm i\Omega_2(\delta, q),$$

$$\kappa_{10,11} = \pm\Omega_3(\delta, q), \qquad \kappa_{12,13} = \pm i\Omega_4(\delta, q), \tag{D.10}$$

and the Rabi frequencies are

$$\Omega_{1,2}(\delta, r) = 2^{-1}[\delta^2 + E_{01}^2(\xi_{8,1}^2 + \xi_{10,1}^2)]^{1/2} = 2^{-1}[\delta^2 + E_{02}^2(\xi_{9,2}^2 + \xi_{11,2}^2)]^{1/2},$$

$$\Omega_{3,4}(\delta, r) = [\delta^2/4 + P/2 \pm (P^2/4 - R)^{1/2}]^{1/2},$$

$$P \equiv 4^{-1}E_{01}^2(\xi_{4,6}^2 + \xi_{4,12}^2 + \xi_{4,13}^2 + \xi_{3,6}^2 + \xi_{3,12}^2 + \xi_{3,13}^2),$$

$$R \equiv 16^{-1} E_{01}^4 [(\xi_{3,6}\xi_{4,12} - \xi_{3,12}\xi_{4,6})^2 + (\xi_{3,6}\xi_{4,13} - \xi_{3,13}\xi_{4,6})^2$$
$$+ (\xi_{3,12}\xi_{4,13} - \xi_{3,13}\xi_{4,12})^2], \quad \xi_{\alpha,\beta} \equiv J_r(\Delta z_{\alpha\beta} E_{02}/\omega_2) z_{\alpha\beta}.$$

$$\tag{D.11}$$

The numeration of the states in (D. 6, 7) is

$$(001) \equiv 1, \quad (00-1) \equiv 2, \quad (100) \equiv 3, \quad (010) \equiv 4,$$
$$(002) \equiv 5, \quad (00-2) \equiv 7, \quad (110) \equiv 6, \quad (101) \equiv 8,$$
$$(10-1) \equiv 9, \quad (011) \equiv 10, \quad (01-1) \equiv 11, \quad (200) \equiv 12, \quad (020) \equiv 13.$$

$$\tag{D.12}$$

The validity of (D.8–11) is controlled by the inequality

$$\Omega^{-1}\Omega_k(0, r) \ll 1 \quad (k = 0, 1, 2, 3, 4). \tag{D.13}$$

Using (3.4.2) and (D.8, 10) we find the quasienergies (with respect to $E_a^{(0)}$)

$$\varepsilon_k^{\pm} = -\delta/2 \pm \Omega_k(\delta, r), \quad (\kappa = \pm i\Omega_k(\delta, r), \quad k = 0, 1, 2, 3, 4),$$
$$\varepsilon = 0, \quad (\kappa = -i\delta/2). \tag{D.14}$$

Now consider the hydrogen atom evolution in a bichromatic EF using the L_α line as an example. We represent the laser field amplitude as $E_{01} = E_x e_1 + E_y e_2 + E_z e_3$, where e_1, e_2, e_3 are the unit vectors directed along the axes $0x, 0y, 0z$. At $\delta = 0$, from (D.8, 9) we get

$$\kappa_1 = i\Omega(r), \quad \kappa_2 = -i\Omega(r), \quad \kappa_{3,4,5} = 0,$$
$$\Omega(r) = \eta R(r), \quad \eta \equiv 2^7 3^{-5},$$
$$R(r) = \{2^{-1}[(E_x^2 + E_y^2)\delta_{0r} + E_z^2 J_r^2(\beta_2)]\}^{1/2}, \tag{D.15}$$

where δ_{0q} is the Kronecker symbol, $\beta_2 \equiv 3E_{02}/\omega_2$. In (D.15) the relations

$$(E_{01}r)_{01} = \eta(E_x + iE_y), \quad (E_{01}r)_{02} = \eta(E_x - iE_y),$$
$$(E_{01}r)_{03} = -\eta E_z, \quad (E_{01}, r)_{04} = \eta E_z, \tag{D.16}$$

are used; $\varphi_1, \varphi_2, \varphi_3, \varphi_4$ are defined in (D.8), and $\varphi_0 \equiv \varphi_{000}$.
The general solution of (D.4) can be written as a superposition of solutions corresponding to the given quasienergies (D.6)

$$X(t) = \sum_{j=1}^{5} C_j X_j(0) \exp(\kappa_j t), \tag{D.17}$$

where C_j are arbitrary constants. The column-vectors $X_j(0)$ have the following form:

Case 1: $(E_x^2 + E_y^2)\delta_{0r} \neq 0$.

$$X_1(0) = \begin{bmatrix} -\dfrac{(E_x - iE_y)\exp(-i\theta)}{2^{3/2}R(r)}\delta_{0r} \\[2mm] -\dfrac{(E_x + iE_y)\exp(-i\theta)}{2^{3/2}R(r)}\delta_{0r} \\[2mm] \dfrac{E_z J_r(\beta_2)\exp(-i\theta)}{2^{3/2}R(r)} \\[2mm] \dfrac{(-1)^{r+1}E_z J_r(\beta_2)\exp(-i\theta)}{2^{3/2}R(r)} \\[2mm] 2^{-1/2} \end{bmatrix},$$

$$X_2(0) = \begin{bmatrix} \dfrac{(E_x - iE_y)\exp(-i\theta)}{2^{3/2}R(r)}\delta_{0r} \\[2mm] \dfrac{(E_x + iE_y)\exp(-i\theta)}{2^{3/2}R(r)}\delta_{0r} \\[2mm] -\dfrac{E_z J_r(\beta_2)\exp(-i\theta)}{2^{3/2}R(r)} \\[2mm] \dfrac{(-1)^r E_z J_r(\beta_2)\exp(-i\theta)}{2^{3/2}R(r)} \\[2mm] 2^{-1/2} \end{bmatrix},$$

$$X_3(0) = \begin{bmatrix} \dfrac{E_x - iE_y}{[2(E_x^2 + E_y^2)]^{1/2}} \\[2mm] \dfrac{-E_x + iE_y}{[2(E_x^2 + E_y^2)]^{1/2}} \\[2mm] 0 \\ 0 \\ 0 \end{bmatrix}, \qquad X_4(0) = \begin{bmatrix} 0 \\ 0 \\ 2^{-1/2} \\ 2^{-1/2} \\ 0 \end{bmatrix},$$

$$X_5(0) = \begin{bmatrix} -\dfrac{E_z J_0(\beta_2)(E_x - iE_y)}{2R(0)(E_x^2 + E_y^2)^{1/2}} \\[2mm] \dfrac{E_z J_0(\beta_2)(E_x + iE_y)}{2R(0)(E_x^2 + E_y^2)^{1/2}} \\[2mm] -\dfrac{(E_x^2 + E_y^2)^{1/2}}{2R(0)} \\[2mm] \dfrac{(E_x^2 + E_y^2)^{1/2}}{2R(0)} \\[2mm] 0 \end{bmatrix}.$$

Case 2: $(E_x^2 + E_y^2)\delta_{0r} = 0, \quad E_z J_r(\beta_2) \neq 0$:

$$X_3(0) = \begin{bmatrix} 1 \\ 0 \\ 0 \\ 0 \\ 0 \end{bmatrix}, \quad X_4(0) = \begin{bmatrix} 0 \\ 1 \\ 0 \\ 0 \\ 0 \end{bmatrix}, \quad X_5(0) = \begin{bmatrix} 0 \\ 0 \\ (-1)^r 2^{-1/2} \\ 2^{-1/2} \\ 0 \end{bmatrix}.$$

the vectors $X_1(0)$ and $X_2(0)$ are the same as in case 1.

Case 3: $(E_x^2 + E_y^2)\delta_{0r} = 0$; $E_z J_r(\beta_2) = 0$:
In this case $\boldsymbol{X}_j \equiv \{X_j(0)\}_i = \delta_{ij}$; i, $j = 1, 2, 3, 4, 5$.

As an example of how to use (B.11–15) we write, for the case of exact resonance, the probabilities $W_{0 \to k}(t)$ of hydrogen atom transitions from the state $\psi_0(t)$ into the state $\psi_j(t)$ $(j = 1, 2, 3, 4)$ under the action of a laser field. Recall that the states $\psi_j(t)$ are determined by the field $\boldsymbol{E}_{02}(t)$. Assuming that at $t = 0$ the atom was in the state $\psi_0(t)$ we obtain

$$W_{0 \to 1}(t) = W_{0 \to 2}(t) = \frac{(E_x^2 + E_y^2)\delta_{0r}\, \sin^2\, \Omega(r)t}{2[(E_x^2 + E_y^2)\delta_{0r} + E_z^2 J_r(\beta_2)]},$$

$$W_{0 \to 3}(t) = W_{0 \to 4}(t) = \frac{E_z^2 J_r^2(\beta_2)\, \sin^2\, \Omega(r)t}{2[(E_x^2 + E_y^2)\delta_{0r} + E_z^2 J_r^2(\beta_2)]},$$

$$W_{0 \to 0}(t) = \cos^2\, \Omega(r)t. \tag{D.18}$$

Equations (D.18) show that the probabilities of populating the excited states $\psi_j(t)$ $(j = 1, 2, 3, 4)$ oscillate at the frequency equal to twice the Rabi frequency $2\Omega(r)$. The amplitude of such oscillations for a given excited state $\psi_j(t)$ is determined by the mutual orientation of the vectors $\boldsymbol{E}_{01}$ and $\boldsymbol{E}_{02}$ as well as by the number r of low-frequency quanta involved into the resonance and the "reduced" amplitude $\beta_2 \equiv 3E_{02}/\omega_2$ of the low-frequency field.

E Influence of Bound Electrons on the Frequency and Damping of Langmuir Oscillations

To estimate the influence of bound electrons we first write the dielectric constant of a plasma as $\varepsilon \approx 1 + 4\pi(\chi_f + \chi_b)$, where $\chi_f = -\omega_{pe}^2/4\pi\omega^2$ is the contribution of the free electrons to the susceptibility, and χ_b is the contribution of the bound electrons [E.1]. In the simplest case of a two-level emitter, at the resonance approximation $(|\omega - \omega_{21}| \ll \omega_{21})$ far from saturation we have [E.1],

$$\varepsilon \approx 1 - \frac{\omega_{pe}^2}{\omega^2} + a\frac{\Gamma}{\omega_{21} - \omega - i\Gamma}, \qquad a \equiv \frac{N_1^{(0)} - N_2^{(0)})|d_{12}|^2}{\hbar\Gamma}, \tag{E.1}$$

where $N_1^{(0)} - N_2^{(0)}$ is the difference of the unperturbed populations. In dense plasmas, electron impact broadening contributes dominantly to the width Γ of a resonance. For hydrogen-like levels, employing the usual estimations [E.2],

$$\Gamma \approx \frac{32}{3}\left(\frac{\hbar}{m_e}\right)^2 \left(\frac{\pi m_e}{8T_e}\right)^{1/2} n^4 N_e \Lambda, \quad \left(\Lambda \equiv \ln \frac{T_e}{n^2 \hbar \omega_{pe}}\right), \quad |d_{12}|^2 \approx \frac{n^4 \hbar^4}{m_e^2 e^2}, \tag{E.2}$$

we find

$$a \approx \frac{20.5}{\Lambda}\left(\frac{T_e}{m_e c^2}\right)^{1/2} \frac{N_1^{(0)} - N_2^{(0)}}{N_e}. \tag{E.3}$$

It can be seen that practically always $a \ll 1$. Indeed, in order for a to become ~ 1 it is necessary to have $(N_1^{(0)} - N_2^{(0)})/N_e \sim (\Lambda/20.5) \cdot (m_e c^2/T_e)^{1/2} \equiv f(T_e)$, where $f(1 \text{ eV}) \approx 300$, $f(10^2 \text{ eV}) \approx 30$. Since the emitter density $N_0 \gg (N_1^{(0)} - N_2^{(0)})$, then from the condition $a \sim 1$ the inequality $N_0/N_e \gg (N_1^{(0)} - N_2^{(0)})/N_e \sim 10^2$ follows which is extremely improbable in practice.

Thus, even a resonance of a Langmuir wave with an atomic transition (inside a multiplet), occurring under the formation of dips on SL profiles, cannot prevent the existence of this wave in a plasma.

References

Chapter 1

1.1 R.Z. Sagdeev, A.A. Galeev: Nonlinear plasma theory, in *Review of Plasma Physics*, Vol. 7, ed. by M.A. Leontovich (Consultants Bureau, New York 1979) p.1
1.2 H.R. Griem: *Spectral Line Broadening by Plasmas* (Academic, New York 1974)
1.3 G.V. Sholin: Sov. Phys. – Dokl. **15**, 1040 (1971)
1.4 A. Boileau, M.V. Hellermann, W. Mandl, H.P. Summers, H. Weisen, A. Zinoviev: J. Phys. B **22**, L 145 (1989)
1.5 V.S. Lisitsa, S.I. Jakovlenko: Sov. Phys. – JETP **39**, 759 (1974); ibid. **41**, 233 (1975)
1.6 Ja.B. Zel'dovich: Sov. Phys. – JETP **24**, 1006 (1967)
1.7 V.I. Ritus: Sov. Phys. – JETP **24**, 1041 (1967)
1.8 I.I. Sobelman, L.A. Vainshtein, E.A. Yukov: *Excitation of Atoms and Broadening of Spectral Lines*, Springer Ser. Chem. Phys., Vol. 7 (Springer, Berlin, Heidelberg 1981) Chap. 7
1.9 E. Oks: in *Spectral Line Shapes*, vol. 7, ed. by R. Stamm and B. Talin (Nova Science Publishers, New York 1993) p. 65
1.10 E.A. Oks: Proc. 14th Summer School and Int' Symp. on Phys. of Ionized Gases (Sarajevo, Yugoslavia 1988) Invited Lectures, p.435
1.11 L.D. Landau, E.M. Lifshitz: *Quantum Mechanics* (Pergamon, Oxford 1965)

Chapter 2

2.1 L.D. Landau, E.M. Lifshitz: *Quantum Mechanics* (Pergamon, Oxford 1965) Chap. 6
2.2 L.I. Schiff: *Quantum Mechanics* (McGraw-Hill, New York 1968) Sect. 35
2.3 A. Messiah: *Quantum Mechanics*, Vol. 2 (North-Holland, Amsterdam 1969) Chap. 17
2.4 I.I. Sobelman: *Atomic Spectra and Radiative Transitions*, 2nd. edn., Springer Ser. Atoms Plasmas, Vol. 12 (Springer, Berlin, Heidelberg 1992) p.
2.5 J.M. Shirley: Phys. Rev. **138**, 979 (1965)
2.6 H. Sambe: Phys. Rev. A **7**, 2203 (1973)
2.7 Ja.B. Zel'dovich: Sov. Phys. – Usp. **16**, 427 (1973)
2.8 I.L. Tomashevsky: Fine structure of hydrogen-like atom quasienergy levels in a varying electric field, in *Spektroskopiya mnogozaryadnikh ionov* [Spectroscopy of Multiply Charged Ions] (USSR Acad. Sci. Scientific Council on Spectroscopy, Moscow 1986) p. 109 (in Russian)

Chapter 3

3.1 D.I. Blochinzew: Phys. Z. Sow. Union **4**, 501 (1933)
3.2 E.V. Lifshitz: Sov. Phys. – JETP **26**, 570 (1968)
3.3 E.A. Oks: Sov. Phys. – Dokl. **29**, 224 (1984)
3.4 E.A. Oks, Yu.M. Shagiev: Tables of Balmer spectral line profiles of hydrogen in a field $E_0 \cos \omega t$. Preprint No. 77, Inst. of Appl. Physics, USSR Acad. Sci., Gor'kii (1983) (in Russian)
3.5 E.A. Oks: Proc. 14th Summer School and Int' Symp. on Phys. of Ionized Gases (Sarajewo, Yugoslavia 1988) p. 313

3.6   I.M. Gaisinsky, E.A. Oks, S.E. Frid: Stark profiles of hydrogen and ionized helium lines in a linearly polarized field $E(t) = \sum_k E_k \cos(\omega t + \varphi_k)$, in *Protzessy vo vnutrennikh atomnikh obolotchkakh* [Processes in Inner Atomic Shells] (USSR Acad. Sci. Scientific Council on Spectroscopy, Moscow 1986) p.75 (in Russian)

3.7   V.P. Gavrilenko, E.A. Oks: Resonances in polychromatic fields containing strong low-frequency components, in Ref. 3.6, p. 213 (in Russian)

3.8   E.V. Aferenko, E.A. Oks: Effect of narrowing of hydrogen-like spectral lines in a quasimonochromatic electric field, in *Mnogochastichnije effekty v atomakh* [Many-Particle Effects in Atoms] (USSR Acad. Sci. Scientific Council on Spectroscopy, Moscow 1985) p. 187 (in Russian)

3.9   V.P. Gavrilenko, E.A. Oks: Proc. 19th Int' Conf. on Phenom. in Ionized Gases (Belgrade, Yugoslavia 1989) p. 354

3.10   M. Abramowitz, I.A. Stegun (eds.): *Handbook of Mathematical Functions with Formulas, Graphs, and Mathematical Tables* (Dover, New York 1964)

3.11   E. Scrödinger: Z. Phys. **78**, 309 (1932)

3.12   Yu.K. Verevkin, A.D. Tertishnik: Sov. J. Plasma Phys. **13**, 287 (1987)

3.13   N.B. Delone, V.A. Kovarskil, A.V. Masalov, N.F. Perelman: Sov. Phys. – Usp. **23**, 472 (1980)

3.14   A.B. Underhill, J.M. Waddell: National Bureau of Standards Circular No. 603, Washington, DC (1959)

3.15   A.V. Galdetsky, E.A. Oks: Bull. Crimean Astrophys. Observ. **65**, 54 (1982)

3.16   V.S. Lisitsa: Sov. Phys. – Usp. **20**, 603 (1977)

3.17   H.R. Griem: Spectral Line Broadening by Plasmas (Acad. Press, New York, London 1974)

3.18   G.V. Sholin, A.V. Demura, V.S. Lisitsa: Sov. Phys. – JETP **37**, 1057 (1973)

3.19   I.I. Sobelman, L.A. Vainshtein, E.A. Yukov: *Excitation of Atoms and Broadening of Spectral Lines*, Springer Ser. Chem. Phys., Vol. 7 (Springer, Berlin, Heidelberg 1981)

3.20   Ya. Ispolatov, E. Oks: Journ. Quant. Spectr. Rad. Transfer **50**, 129 (1993)

3.21   V.S. Butylkin, A.E. Kaplan, Yu.G. Khronopulo, Yakubovich: *Resonant Nonlinear Interactions of Light with Matter* (Springer, Berlin, Heidelberg 1989)

3.22   R.H. Pantell, H.E. Puthoff: *Fundamentals of Quantum Electronics* (Wiley, New York 1969)

3.23   V.A. Kovarsky: *Multiquantum Transitions* (Shtiintza, Kishinev 1974)

3.24   V.P. Gavrilenko, E.A. Oks: Sov. Tech. Phys. Lett. **10**, 609 (1984)

Chapter 4

4.1   E.A. Oks, V.P. Gavrilenko: Opt. Commun. **46**, 205 (1983)

4.2   V.P. Gavrilenko, E.A. Oks: Sov. Phys. – JETP **53**, 1122 (1981)

4.3   V.P. Gavrilenko, E.A. Oks: Sov. J. Plasma Phys. **13**, 22 (1987)

4.4   V.P. Gavrilenko, E.A. Oks: Proc. 17th Int' Conf. on Phenom. in Ionized Gases (Budapest, Hungary 1985) p. 1081

4.5   V.P. Gavrilenko, B.B. Nadezhdin, E.A. Oks: Proc. 8th Europ. Sect. Conf. on Atomic and Molecular Physics of Ionized Gases (Greifswald, GDR 1986) p. 132

4.6   V.P. Gavrilenko, B.B: Nadezhdin, E.A. Oks: Proc. 18th Int' Conf. on Phenom. in Ionized Gases (Swansea, UK 1987) p. 462

4.7   D.A. Volod'ko, V.P. Gavrilenko: Opt. Spectrosc. **64**, 155 (1988)

4.8   T. Ishimura: J. Phys. Soc. Jpn. **23**, 422 (1967)

4.9   V.S. Lisitsa: Opt. Spectrosc. **31**, 468 (1971)

4.10   V.S. Lisitsa: Sov. Phys. – Usp. **20**, 603 (1977)

4.11   L.D. Landau, E.M. Lifshitz: *Quantum Mechanics* (Pergamon, Oxford 1965)

4.12   M.D. Anosov: Opt. Spectrosc. **47**, 121 (1979)

4.13   I.I. Rabi: Phys. Rev. **51**, 652 (1932)

4.14   D.I. Blochinzew: Phys. Z. Sow. Union **4**, 501 (1933)

4.15   V.V. Alikaev: Proc. 10th Europ. Conf. on Controlled Fusion and Plasma Phys., Vol. 2 (Moscow, USSR 1981) p. 11

4.16   N.N. Bogoliubov, Yu.M. Mitropolskii: *Asymptotic Methods in the Theory of Nonlinear Oscillations* (Gordon and Breach, New York 1961)
4.17   S.H. Kim, H.E. Wilhelm: J. Appl. Phys. **44**, 802 (1973)
4.18   W.R. Rutgers, H.W. Kalfsbeek: Z. Naturforsch. **30a**, 739 (1975)
4.19   A. Cohn, P. Bakshi, G. Kalman: Phys. Rev. Lett. **29**, 34 (1972) [Corrigenda: Phys. Rev. Lett. **31**, 620 (1973)]
4.20   C.C. Gallagher, M.A. Levine: Phys. Rev. Lett. **30**, 897 (1973)
4.21   W.R. Rutgers, H. de Kluiver: Z. Naturforsch. **29a**, 42 (1974)
4.22   V.P. Gavrilenko: Sov. Phys. – JETP **67**, 915 (1988)
4.23   P. Bakshi, G. Kalman, A. Cohn: Phys. Rev. Lett. **31**, 1576 (1973)
4.24   H.R. Griem: in *Spectral Line Shapes*, vol. 7, ed. by R. Stamm and B. Talin (Nova Science Publishers, New York 1993) p. 3
4.25   S. Günter, A. Könies: Phys. Rev. E **49**, 4732 (1994)
4.26   D.V. Fursa, G.L. Yudin: Phys. Rev. A **44**, 7414 (1991)
4.27   E.A. Oks, St. Böddeker, H.-J. Kunze: Phys. Rev. A **44**, 8338 (1991)
4.28   G.V. Sholin, V.S. Lisitsa, V.I. Kogan: Sov. Phys. JETP **32**, 758 (1971)

Chapter 5

5.1   M. Baranger, B. Mozer: Phys. Rev. **123**, 25 (1961)
5.2   W.S. Cooper, H. Ringler: Phys. Rev. **179**, 226 (1969)
5.3   E.A. Oks, V.P. Gavrilenko: Sov. Tech. Phys. Lett. **9**, 111 (1983)
5.4   M.P. Brizhinev, V.P. Gavrilenko, S.V. Egorov, B.G. Eremin, A.V. Kostrov, E.A. Oks, Yu.M. Shagiev: Sov. Phys. – JETP **58**, 517 (1983)
5.5   N.F. Perelman, A.A. Mosyak: Sov. Phys. – JETP **69**, 700 (1989)
5.6   V.P. Gavrilenko, E.A. Oks: Proc. Int' Conf. on Plasma Physics (Göteborg, Sweden 1982) p. 353
5.7   I.M. Gaisinsky, E.A. Oks: A new effect of spectral line shift under interaction of laser radiation with a plasma, in *Korrelatzionniye i relativistskiye effekty v atomakh i ionakh* [Correlation and Relativistic Effects in Atoms and Ions] (USSR Acad. Sci. Scientific Council on Spectroscopy, Moscow 1986) p. 106 (in Russian)
5.8   V.P. Gavrilenko, E.A. Oks: Sov. J. Quantum Electron. **13**, 1269 (1983)
5.9   V.P. Gavrilenko, E.A. Oks: Opt. Commun. **69**, 384 (1989)
5.10   V.P. Gavrilenko, E.A. Oks: Proc. 13th Summer School and Int' Symp. on Phys. of Ionized Gases (Sibenik, Yugoslavia 1986) p. 393
5.11   V.P. Gavrilenko, E.A. Oks: Resonances in polychromatic fields containing strong low-frequency components, in *Protzessy vo vnutrennikh atomnikh obolotchkakh* [Processes in Inner Atomic Shells] (USSR Acad. Sci. Scientific Council on Spectroscopy, Moscow 1986) p. 213 (in Russian)
5.12   H.J. Kunze, H.R. Griem: Phys. Rev. Lett. **21**, 1048 (1968)
5.13   A.S. Davydov: *Quantum Mechanics* (Pergamon, Oxford 1965) Sect. 49
5.14   W.W. Hicks, R.A. Hess, W.S. Cooper: Phys. Rev. A **5**, 490 (1972)
5.15   W.C. Martin: J. Phys. Chem. Ref. Data **23**, 257 (1973)
5.16   K. Kumar, N. Peitel, N. Bloembergen: In the World of Science **11**, 4 (1987) (in Russian)
5.17   D. Levron, G. Benford, D. Tzach: Phys. Rev. Lett. **58**, 1336 (1978)
5.18   I.I. Sobelman: *Atomic Spectra and Radiative Transitions*, 2nd edn., Springer Ser. Atoms Plasmas, Vol. 12 (Springer, Berlin, Heidelberg 1992)
5.19   E.A. Oks, G.V. Sholin: Sov. Phys. – JETP **41**, 482 (1975)
5.20   I.I. Rabi: Phys. Rev. **51**, 652 (1932)
5.21   N.N. Bogoliubov, Yu.M. Mitropolskii: *Asymptotic Methods in the Theory of Nonlinear Oscillations* (Gordon and Breach, New York 1961)
5.22   H.A. Bethe, E.E. Salpeter: *Quantum Mechanics of One- and Two-Electron Atoms* (Plenum, New York 1977)
5.23   D.I. Blochinzew: Phys. Z. Sow. Union **4**, 501 (1933)

5.24   H.R. Griem: *Spectral Line Broadening by Plasmas* (Academic, New York 1974)
5.25   H.R. Griem: Phys. Rev. A **27**, 2566 (1983)
5.26   G.A. Moore, G.P. Davis, R.A. Gottscho: Phys. Rev. Lett. **52**, 538 (1984)

Chapter 6

6.1   D.A. Volod'ko, V.P. Gavrilenko, E.O. Oks: Proc. 18th Int' Conf. on Phenom. in Ionized Gases (Swansea, UK 1987) p. 604
6.2   E.A. Oks, V.P. Gavrilenko: Opt. Commun. **56**, 415 (1986)
6.3   D.A: Volod'ko, V.P. Gavrilenko, E.A. Oks: In *Spectral Line Shapes*, Vol. 5 9th Int' Conf. (Torun, Poland 1988) p. A28
6.4   V.P. Gavrilenko, B.B: Nadezhin, E.A. Oks: Proc. 18th Int' Conf. on Phenom. in Ionized Gases (Swansea, UK 1987) p. 462
6.5   B.B. Nadezhdin, E.A. Oks: Proc. 8th Europ. Sect. Conf. on Atomic and Molecular Physics of Ionized Gases (Greifswald, GDR 1986) p. 132; Proc. 13th Summer School and Int' Symp. on Phys. of Ionized Gases (Sibenik, Yugoslavia 1986) p. 397
6.6   W.S. Cooper, H. Ringler: Phys. Rev. **179**, 226 (1969)
6.7   J. Holtsmark: Ann. Phys. **58**, 577 (1919)
6.8   P.A. Braun, A.N. Petelin: Sov. Phys. – JETP **39**, 775 (1974)
6.9   L.D. Landau, E.M. Lifshitz: *Mechanics* (Pergamon, Oxford 1960)
6.10   A.I. Baz', Ya.B. Zeldovich, A.M. Perelomov: *Scattering, Reactions and Decays in Nonrelativistic Quantum Mechanics* (Israel Program for Scientific Translations, Jerusalem 1969)

Chapter 7

7.1   A.B. Underhill, J.H. Waddell: National Bureau of Standards Circular No. 603, Washington, DC (1959)
7.2   M.V. Babykin, A.I. Zhuzhunashvili, E.A. Oks, V.V. Shapkin, G.V. Sholin: Sov. Phys. – JETP **38**, 86 (1974)
7.3   A.I. Zhuzhunashvili, E.A. Oks: Sov. Phys. – JETP **46**, 1122 (1977)
7.4   E.A. Oks, G.V. Sholin: Opt. Spectrosc. **42**, 434 (1977)
7.5   A.B. Berezin, B.V. Ljublin, D.G. Jakovlev: Sov. Phys. – Tech. Phys. **28**, 407 (1983)
7.6   E.A. Oks, V.A. Rantsev-Kartinov: Sov. Phys. – JETP **52**, 50 (1980)
7.7   Nguen-Hoe, H.W. Drawin, L. Herman: J. Quant. Spectrosc. Radiat. Transfer **7**, 429 (1967)
7.8   A. Piel, H. Richter: Z. Naturforsch. **34 a**, 516 (1979)
7.9   H.M. Crosswhite: *The Hydrogen Molecular Wavelength Tables of G.H. Dieke* (Wiley-Interscience, New York 1972)
7.10   H.W. Drawin, J. Ramette: Z. Naturforsch. **33a**, 1285 (1978)
7.11   F. Pinnekamp: Z. Naturforsch. **34a**, 529 (1979)
7.12   E.A. Oks, V.A. Rantsev-Kartinov: Preprint No. 3161 of the Inst. of Atomic Energy, Moscow (1979)
7.13   E.A. Oks, G.V. Sholin: Sov. Phys. – Tech. Phys. **21**, 144 (1977)
7.14   V.V. Alexandrov, A.I. Gorlanov, N.G. Koval'skii, S.Yu. Luk'yanov, V.A. Rantsev-Kartinov: In *Diagnostika plazmy [Diagnostics of Plasmas]*, No. 3 (Atomizdat, Moscow 1973) pp. 79, 200 (in Russian
7.15   E.A. Oks, G.V. Sholin: Sov. Phys. – JETP **41**, 482 (1975)
7.16   S.K. Zhdanov, B.A. Trubnikov: Sov. Phys. – JETP **56**, 1197 (1982)
7.17   K.H. Finken, R. Buchwald, G. Bertschinger H.-J. Kunze: Phys. Rev. A **21**, 200 (1980); K.H. Finken: Fortschr. Phys. **31**, 1 (1983)
7.18   G. Bertschinger: "Messungen von VUV Linien an einem dichten Z-Pinch-Plasma"; Ph. D. Thesis, Ruhr-University, Bochum (1980)
7.19   E.A. Oks, St. Böddeker, H.-J. Kunze: Phys. Rev. **A44**, 8338 (1991)
7.20   B. Yaakobi, D. Steel, E. Thorsos, A. Hauer, B. Perry: Phys. Rev. Lett. **39**, 1526 (1977)

7.21 A.A. Bagdasarov, V.I. Bugarya, N.L. Vasin, V.A. Vershkov: Proc. 12th Europ. Conf. on Controlled Fusion and Plasma Phys. (Budapest, Hungary 1985) p.207
7.22 V.P. Gavrilenko, E.A. Oks, V.A. Rantsev-Kartinov: JETP Lett. **44**, 404 (1986)
7.23 E. Dullni, P. Leismann, S. Maurmann, C.V. Reventlow, H.-J. Kunze: Phys. Scr. **34**, 405 (1986)
7.24 A. Boileau, M.V. Hellerman, W. Mandl, H.P. Summers, H. Weisen, A. Zinoviev: J. Phys. B **22**, L 145 (1989)
7.25 V.A. Abramov, V.S. Lisitsa: Sov. J. Plasma Phys. **3**, 451 (1977)
7.26 S.S. Bychkov, R.S. Ivanov, G.I. Stotskii: Sov. J. Plasma Phys. **13**, 769 (1987)
7.27 A. Kamp, G. Himmel: Appl. Phys. B **47**, 177 (1988)
7.28 R.A. Akhmedzhanov, I.N. Polushkin, Yu.V. Rostovtsev, M.Yu. Ryabikin, Yu.M. Shagiev, V.V. Yazenkov: Sov. Phys. – JETP **63**, 30 (1986)
7.29 V.M. Baev, T.P. Belikova, E.A. Sviridenkov, A.F. Suchkov: Sov. Phys. – JETP **47**, 21 (1978); T.P. Belikova, E.A. Sviridenkov, A.F. Suchkov: Sov. J. Quantum Electron. **4**, 454 (1974)
7.30 R.E. Shefer, G. Bekefi: Phys. Fluids **22**, 1584 (1979)
7.31 M.P. Brizhinev, S.V. Egorov, B.G. Eremin, A.V. Kostrov, E.A. Oks, Yu.M. Shagiev: Proc. 15th Int' Conf. on Phenom. in Ionized Gases (Minsk, USSR 1981) p. 971
7.32 V.N. Kulikov, V.E. Mitsuk: Sov. Tech. Phys. Lett. **14**, 104 (1988)
7.33 M.P. Brizhinev, V.P. Gavrilenko, S.V. Egorov, B.G. Eremin, A.V. Kostrov, E.A. Oks, Yu.M. Shagiev: Sov. Phys. – JETP **58**, 517 (1983)
7.34 G.V. Zelenin, A.A. Kutsyn, M.E. Maznichenko, O.S. Pavlichenko, V.A. Suprunenko: Sov. Phys. – JETP **31**, 1009 (1970)
7.35 W.S. Cooper, H. Ringler: Phys. Rev. **179**, 226 (1969)
7.36 B.G. Eremin, A.V. Kostrov, A.D. Stephanushkin: Sov. J. Plasma Phys. **5**, 661 (1979)
7.37 A.L. Vicharev, O.A. Ivanov, A.N. Stepanov: Sov. J. Plasma Phys. **14**, 32 (1988)
7.38 I.N. Polushkin, M.Yu. Ryabikin, Yu.M. Shagiev, V.V. Yazenkov: Sov. Phys. – JETP **62**, 953 (1985)
7.39 V.N. Aleinikov, B.G. Eremin, G.L. Klimchitskaya, I.N. Polushkin, Yu.V. Rostovtsev, V.V. Yazenkov: Sov. Phys. – JETP **67**, 908 (1988)
7.40 U. Rebhan, N.J. Wiegart, H.-J. Kunze: Phys. Lett. **85 A**, 228 (1981)
7.41 U. Rebhan: J. Phys. B **19**, 3487 (1986)
7.42 J. Hildebrandt: J. Phys. B **16**, 149 (1983)
7.43 J. Hildebrandt: Opt. Commun. **53**, 229 (1985)
7.44 J. Hildebrandt: J. Quant. Spectrosc. Radiat. Transfer **37**, 211 (1987)
7.45 H.R. Griem: *Spectral Line Broadening by Plasmas* (Academic, New York 1974)
7.46 G. Bekefi, C. Deutsch: Comments Plasma Phys. **2**, 89 (1976)
7.47 H.-J. Kunze: in *Spectral Line Shapes,* ed. by B. Wende (de Gruyter, Berlin 1981) p. 517
7.48 K. Kawasaki, K. Takiyama, T. Oda: Jpn. J. Appl. Phys. **27**, 83 (1988)
7.49 K. Takiyama, Y. Kamiura, T. Fujita, T. Oda, H. Sakai, K. Kawasaki: Jpn. J. Appl. Phys. **26**, 1945 (1987)
7.50 K. Danzmann, K. Grützmacher, B. Wende: Phys. Rev. Lett. **57**, 2151 (1986)
7.51 A. Derevianko, E. Oks: Phys. Rev. Lett. **73**, 2059 (1994)
7.52 V.S. Lisitsa: Sov. Phys. Usp. **20**, 603 (1977)

Appendices

A.1 D.I. Blochinzew: Phys. Z. Sow. Union **4**, 501 (1933)
C.1 V.M. Fain, Ya.I. Khanin: *Quantum Electronics*, vol. I (Pergamon, Oxford 1969)
C.2 V.S. Lisitsa: Sov. Phys. Usp. **20**, 603 (1977)

Subject Index

Printing: Mercedesdruck, Berlin
Binding: Buchbinderei Lüderitz & Bauer, Berlin